CELL BIOLOGY RESEARCH PROGRESS

POTASSIUM CHANNELS

TYPES, STRUCTURE AND BLOCKERS

CELL BIOLOGY RESEARCH PROGRESS

Additional books in this series can be found on Nova's website under the Series tab.

Additional E-books in this series can be found on Nova's website under the E-book tab.

CELL BIOLOGY RESEARCH PROGRESS

POTASSIUM CHANNELS

TYPES, STRUCTURE AND BLOCKERS

DANIELLE S. FONSECA
EDITOR

Nova Biomedical Books
New York

Library of Congress Cataloging-in-Publication Data

Potassium channels : types, structure, and blockers / editor, Danielle S. Fonseca.
p. ; cm.
Includes bibliographical references and index.
ISBN 978-1-61324-880-5 (hardcover : alk. paper) 1. Potassium channels. I. Fonseca, Danielle S.
[DNLM: 1. Potassium Channels--physiology. 2. Potassium Channel Blockers. QU 55.7]
QP535.K1P663 2011
547'.05383--dc23

2011018638

Published by Nova Science Publishers, Inc. † New York

Contents

Preface

Potassium channels are the most widely distributed type of ion channel and are found in virtually all living organisms. They form potassium-selective pores that span cell membranes, and are found in most cell types and control a wide variety of cell functions. This book presents current research in the study of potassium channels, including potassium channel activity in the regulation of vascular tone in physiological and pathophysiological conditions; the contribution of lymphocyte potassium channels in the perinatal regulation of the immune response in both mother and the newborn; slow potassium channel dysfunction in amyotrophic lateral sclerosis and plant mitochondrial potassium channels.

Chapter 1 - One of the main mechanisms regulating the vascular tone in physiological and pathophysiological conditions is the potassium channels activity. These channels are usually predominant in vascular smooth muscle cell membranes and play a major role in the regulation of the membrane potential. Their contribution in the regulation of the membrane potential and in the vessel contractility depends on the vascular bed and on the species. The activation of these channels in smooth muscle cells leads to hyperpolarization and vasorelaxation, whereas their inhibition induces depolarization and contraction. At present, four distinct types of potassium channels have been identified in the vascular smooth muscle cells: voltage-dependent K^+ (K_V) channels, calcium-activated K^+ (K_{Ca}) channels, ATP-sensitive K^+ (K_{ATP}) channels and inward rectifier K^+ (Kir) channels. The Kv channels are activated by depolarization and may contribute to steady state resting membrane potential. The K_{Ca} channels are activated both by elevated concentrations of intracellular calcium and by membrane depolarization. These channels can be divided into three groups: large (BK_{Ca}), intermediate (IK_{Ca}) and small (SK_{Ca})

conductance. The BK_{Ca} are, highly expressed in vascular smooth muscle cells and contribute to control of vascular tone by participating in both vasodilation and vasoconstriction. The K_{ATP} channels are activated by pharmacological and endogenous vasodilators and its activation is also associated with several pathophysiologies. These channels may contribute to maintain the resting membrane potential and the resting tone. The Kir channels are activated by slight changes in extracellular K^+ and may contribute to resting membrane potential and resting tone.

This review describes the basic properties of the several types of potassium channels present in the vascular smooth muscle cells, including the structure, the cellular mechanism regulating its activity and the new advances in the development of activators and blockers of these channels.

Chapter 2 - Voltage-gated Kv1.3 potassium channels and calcium-dependent IKCa1 potassium channels play an important role in the regulation of lymphocyte activation. Upon antigen presentation, they maintain the electrochemical driving force for sustained calcium influx that regulates cytokine production and further components of an adequate immune response.

Maternal and neonatal immune functions are characterized by distinct alterations in the perinatal period compared with the adult, non-pregnant immune status. These physiological characteristics play an important role in human reproduction and early postnatal development. At the same time, pathologic alterations of the immune response may occur both in the maternal and neonatal immune system, resulting in specific diseases. Recent studies demonstrated that lymphocyte potassium channels may be important elements in the development of the physiological and pathologic alterations of the immune system in human pregnancy and the neonatal period.

In healthy pregnancy, the maternal immune system acquires tolerance in order to protect the developing fetus from harmful immunological reactions. If this tolerance is impaired, an uncontrolled maternal immune response may arise, contributing to the development of a pregnancy specific syndrome, preeclampsia. There is a characteristic pattern of lymphocyte calcium influx and activation in healthy pregnancy influenced by potassium channels. This pattern is missing in preeclampsia, where the above properties are rather comparable to the non-pregnant state.

Decreased functionality of neonatal lymphocytes is a widely recognized experimental and clinical phenomenon. Cytokine production in activated T lymphocytes of the term neonate is reduced compared to adults. A possible contributing factor to this reduction might be the impairment of mechanisms regulating short-term activation of lymphocytes compared to adults. Kv1.3 and

IKCa1 lymphocyte potassium channels are essential components of these mechanisms. Recent results indicate that characteristics of short-term activation of major neonatal lymphocyte subsets are indeed altered compared to adults. These findings improve the understanding of the mechanisms that prevent neonatal lymphocytes from adequate activation upon activating stimuli and, hence, exert a loourr intensity of immune response. They show that the functional impairment of lymphocyte potassium channels may be of importance in those mechanisms.

In this chapter, the authors review recent knowledge on the contribution of lymphocyte potassium channels to perinatal immunity both from a maternal and a neonatal perspective. Furthermore, the authors provide data on the role of these channels in shaping the kinetics of calcium influx during lymphocyte activation in pregnancy and in neonates.

Chapter 3 - Cellular homeostasis is fundamental to maintain the cell alive and in this context ions play an important role. Ion movement is tightly controlled by a variety of channels that function as gateways into the cell. All channels are selective for the type of ion that can pass through its pore. The opening of these channels is tightly regulated by several stimuli, such as voltage, pH or the binding of specific molecules. Definite actions on the cellular metabolism are observed once the increase of ions takes place in the cytoplasm. Then, the movement of ions is related to cell signaling acting as second messengers.

Potassium channels are the most studied. In general, the main strategy to known their structural composition and functions has been made by mutations. Similar experiments contributed to know the composition of channels as well as to identify the channel gate. More than 75 different mammalian genes coding for the potassium channels have been described. The channel diversity can be explained as the result of an alternative splicing during mRNA processing. There must be more functional channels than those codified in the genes, since different subunits of the same family can co-assemble to form diverse functional channels.

Chapter 4 - Amyotrophic lateral sclerosis (ALS) is a neurodegenerative disease that predominantly affects upper and lower motor neuron systems. In addition to muscle wasting and weakness, fasciculation is a characteristic clinical and electrophysiological feature in ALS that suggests abnormal excitability of motor nerves. Nerve hyperexcitability itself is neurotoxic, because the reverse action of calcium-sodium exchanger increases the intracellular calcium concentration, that leads to cell death. Recently, ion channel dysfunction has been reported in sporadic and familial ALS patients and model animals, such as smaller slow potassium current and greater persistent sodium current, both of which contribute to the increased nerve excitability. Threshold tracking is a non-invasive in vivo neurophysiologic test that can analyze such ion channel functions. In this chapter, the method of threshold tracking and its findings are discussed. A new class of antiepileptic agent, a slow potassium channel opener, such as retigabine, demonstrated a hyperpolarizing stabilizing effect of the membrane potential. Thus, the slow potassium channel may become a promising target for neurodegenerative diseases by its neuroprotective effect.

Chapter 5 - The history of plant mitochondrial potassium channel/s began about ten years ago, when the first channel was described on a functional basis in durum wheat mitochondria.

This channel was named Plant Mitochondrial Potassium Channel ATP sensitive (PmitoKATP) in analogy with the animal counterpart (mitoKATP). The PmitoKATP shows interesting features, being able to deeply affect mitochondrial bioenergetics and to control mitochondrial reactive oxygen species (ROS) production, thus playing a role as a mechanism acting against oxidative/environmental stresses in mitochondria/cell/plant.

To date, mitochondrial potassium channels have been also described in about ten plant species. However, these channels display at the same time analogies and differences with respect to the original PmitoKATP. The intrinsic activity may vary significantly among channels, the pattern of modulators is different and even some of these channels are ATP-insensitive.

Finally, different physiological roles have been proposed. Awaiting for the identification of the molecular nature of each channel, here the authors point out the question whether the observed differences may be attributed to the same channel differently modulated in mitochondria from different sources or it may be more appropriate to consider different channels. At the moment, it appears more appropriate to refer to different channels, thus suggesting that a fit nomenclature should be consistently used.

Chapter 6 - Fasting improves contractile recovery and attenuates lactate production and mitochondrial permeability transition (MPT) without altering cell viability in ischemic-reperfused rat hearts. 5-hydroxydecanoate (5-HD), a mitochondrial ATP-sensitive K^+ channel (K-ATP_{mito}) blocker, abolishes the improvement of mechanical recovery elicited by fasting despite it decreases lactate production, suggesting that K-ATP_{mito} might be involved in the preservation of contractility. This chapter assessed the contribution of K-ATP_{mito} in the attenuation of MPT which in turn may improve mitochondrial energetics.

Langendorff-perfused hearts from fed and fasted rats were subjected to ischemia-reperfusion in the presence or absence of 5-HD. To assess whether 5-HD has any direct effect on glycolysis, a cell free heart extract containing all the glycolytic enzymes was used. MPT was quantitated measuring the mitochondrial 2-[^{3}H]-deoxyglucose (^{3}H-2-DG) entrapment. Total heart ^{3}H-2-DG content as an estimation of necrosis was measured. As an index of mitochondrial energetic function, the rate of ATP synthesis was examined in mitochondria isolated from the ischemic-reperfused hearts.

Fasting increased the rate of mitochondrial ATP synthesis and attenuated MPT. 5-HD abolished these protective effects. 5-HD did not change total heart ^{3}H-2-DG content indicating that it lacked effects on cell survival. Since 5-HD did not affect glucose consumption and lactate production in the cell-free heart-extract the inhibition of glycolysis could be due to its own oxidative metabolism.

It may be concluded that cardioprotection elicited by fasting in the ischemic-reperfused heart could be ascribed, at least in part, to the K-ATP$_{mito}$ activation which in turn inhibits MPT and induces energetic preservation.

In: Potassium Channels
Editor: Danielle S. Fonseca

ISBN: 978-1-61324-880-5

Chapter I

Potassium Channels and Vascular Tone: Structure, Regulation and Function

Elisa Cairrão* and Ignacio Verde*

*CICS-UBI - Centro de Investigação em Ciências da Saúde, Universidade da Beira Interior, 6200-506 Covilhã, Portugal

Abstract

One of the main mechanisms regulating the vascular tone in physiological and pathophysiological conditions is the potassium channels activity. These channels are usually predominant in vascular smooth muscle cell membranes and play a major role in the regulation of the membrane potential. Their contribution in the regulation of the membrane potential and in the vessel contractility depends on the vascular bed and on the species. The activation of these channels in smooth muscle cells leads to hyperpolarization and vasorelaxation, whereas their inhibition induces depolarization and contraction. At present, four distinct types of potassium channels have been identified in the vascular smooth muscle cells: voltage-dependent K^+ (K_V) channels,

* Dr. Ignacio Verde, Centro de Investigação em Ciências da Saúde, Universidade da Beira Interior. Av. Infante D. Henrique. 6200-506 Covilhã. Portugal. Tel.: +351-275-329049. Fax: +351-275-329099. E-mail: iverde@fcsaude.ubi.pt

calcium-activated K^+ (K_{Ca}) channels, ATP-sensitive K^+ (K_{ATP}) channels and inward rectifier K^+ (Kir) channels. The Kv channels are activated by depolarization and may contribute to steady state resting membrane potential. The K_{Ca} channels are activated both by elevated concentrations of intracellular calcium and by membrane depolarization. These channels can be divided into three groups: large (BK_{Ca}), intermediate (IK_{Ca}) and small (SK_{Ca}) conductance. The BK_{Ca} are, highly expressed in vascular smooth muscle cells and contribute to control of vascular tone by participating in both vasodilation and vasoconstriction. The K_{ATP} channels are activated by pharmacological and endogenous vasodilators and its activation is also associated with several pathophysiologies. These channels may contribute to maintain the resting membrane potential and the resting tone. The Kir channels are activated by slight changes in extracellular K^+ and may contribute to resting membrane potential and resting tone.

This review describes the basic properties of the several types of potassium channels present in the vascular smooth muscle cells, including the structure, the cellular mechanism regulating its activity and the new advances in the development of activators and blockers of these channels.

1. Introdution

The potassium channels are predominant in vascular smooth muscle cells (SMC) and play a major role in the regulation of membrane potential. In order to understand the role of K^+ channels at the arterial smooth muscle, it is crucial to know the physiological range of membrane potentials in the SMC [1]. Depending of type of blood vessel, the vascular SMC have a resting membrane potential between -40 and -60mV, [2-5]. Therefore, the permeability of K^+ does not completely dominate the membrane potential conductance, once the membrane potential is considerably more positive than the E_K (about -85mV). The most probably reason is that the Cl^- conductance is relatively high, about -31mV [4], and the intracellular concentration of this ion in the vascular SMC is high [6, 7], resulting in a more positive resting membrane potential. When the E_K is substantially more negative that the resting membrane potential, the opening of K^+ channels may induce hyperpolarization and, at the contrary, the closing of K^+ channels may cause depolarization. Generally, the opening of K^+ channels induces the efflux of K^+ for the extracellular medium, which may cause hyperpolarization or repolarization of the cell membrane, closure of the voltage-dependent Ca^{2+}-

channels, decrease in calcium entry into the cell and vasodilatation. On the contrary, the closure of the K^+ channels causes the opening of the voltage-dependent Ca^{2+}-channels, increase of intracellular calcium and vasoconstriction [1, 5, 8, 9].

In vascular SMC, at present, four distinct types of K^+ channels have been identified: the voltage-dependent K^+ (K_V) channels, the calcium-activated K^+ (K_{Ca}) channels, the ATP-sensitive K^+ (K_{ATP}) channels and the inward rectifier K^+ (Kir) channels. The presence of these channels varies and the distribution in the cell is not homogeneous. For example, there are 100 to 500 K_{ATP} and Kir channels per cell, and 1000 to 10000 Kv and K_{Ca} channesl per cell. Moreover, the current intensity of these different K^+ channels, measured by the patch-clamp technique, is very different [1].This review describes the basic properties of these types of vascular K^+ channels, analyzing the structure, the cellular mechanism regulating its activity and the new advances in the development of regulators of these channels.

2. Voltage-Dependent K^+ Channels (K_V)

The SMC K_V channels are activated by despolarization. A small depolarization in vascular SMC also activates L-type voltage-dependent Ca^{2+} channels that induced the Ca^{2+} influx and stimulation of contractile apparatus; for that reason the activity of K_V channels is important for regulating cell excitability and maintaining basal tone [1, 10-13].

On the basis of voltage dependence, pharmacology, activation, inactivation and kinetics, two types of K_V current were identified in vascular SMC: delayed rectifier outward K^+ current and transient outward K^+ current, [1, 12]. The delayed rectifier outward K^+ current has fast activation and slow inactivation, whereas in the transient outward K^+ current both activation and inactivation are rapid [12, 14].

The delayed rectifier outward K^+ current is present in most of vascular SMC, whereas transient outward K^+ current is less frequently detected in these cells. The two types of currents can also coexist, for example, in renal resistance arteries [15] and pulmonary artery [16]. In general, the studies using the whole cell configuration of the Patch clamp technique were not able to differentiate these two types of currents. At present, there are not drugs allowing to discriminate these two currents and the kinetics of both are quite difficult to analyzed when different types of K^+ channels are expressed [14, 17-20].

Structure of K_V Channels

The channels are structurally similar to the first cloned K_V channel, the *Drosophila* Shaker channel [21] All mammalian K_V channels consist of four α-subunits, each containing six transmembrane α-helical segments S1–S6 and a membrane-reentering P-loop (P), which are arranged circumferentially around a central pore as homo- or heterotetramers [22]. This ion-conduction pore is lined by four S5-P-S6 sequences while the four S1–S4 segments, each containing four positively charged arginine residues in the S4 helix, that act as voltage-sensor domains and "gate" the pore by "pulling" on the S4–S5 linker [22-24]. The landmark in the understanding of the molecular basis for channel function was the publication of the K^+ channel structure in 1998 [25], which revealed the overall architecture of the channel and illuminated the basis of the channel selectivity for K^+ over other cations. The subject of that study was the pH-activated prokaryotic channel KcsA. Despite the determination of structures of other channel classes, and the extensive supportt of biophysical data, the means by which K^+ conduction is switched on and off has remained elusive [26]. The ion conduction pore in the K_V channels consists in a sequence of approximately 20 amino acids (P-region) between the S5 and S6 segments, with contributions of S6 and the S4–S5 linker. The subunits are oriented such that the S5–P–S6 sections face each other creating the central pore [27, 28]. The pore is constituted by an inverted cone (teepee), with the selectivity filter apprehended at its wide end. This filter is disposed in an optimal geometry so that a dehydrated K^+ ion fits with proper coordination but the Na^+ ion is too small, and this explains the fact that these channels are 10,000 times more permeable to K^+ than Na^+, even if the radius of K^+ is bigger that of Na^+ (1.33 and 0.95A respectively) [25].

The voltage-sensor domains (VSD) are positioned at the periphery of the channel and consist of four transmembrane segments (S1-S4). Structural rearrangement of the voltage-sensor domains in response to changes in the membrane potential, and in particular S4, which includes positively charged amino acids at every third position, results in conformational changes in the conduction pore, which could open or occlude the ion conduction pathway [27, 29]. Moreover, a large number of recent studies indicates that the first four arginine residues in S4 contributes with 12–13 electronic charges per channel and there are translocated across the membrane's electric field [27, 30]. Although it is well accepted that the movement of the voltage-sensing S4 helix is tightly coupled to opening and closing of the cytoplasmic S6 channel gate; however, the nature of S4 motion is uncertain. Specifically, the topology of the VSD in the closed state and the magnitude of the S4 movement following depolarization remain controversial. So far, three main models of VSD motion have been proposed: (i) the transporter model, in which S4 moves only a small distance (2–3 Å), but through a focused and mobile electric field within an aqueous crevice whose accessibility changes during gating [31]; (ii) the helical screw model, in which the S4 helix rotates clockwise and translates outward (~13 Å) along its axis to move the gating charges across the membrane electric field [32]; (iii) the paddle model where the sensing unit (S4-S3b) undergoes a large transverse movement (~15–20 Å) across the membrane and in which the S4 arginines are mostly exposed to lipids [30, 33]. Additional structural studies will be necessary to clarify the nature of these movements and conformational changes in the voltage-sensors, and the precession mechanisms that are involved in the modulation of these channels. More than 70 genes have been identified in the human genome that code for members of the voltage gated K^+ channel superfamily [34]. These include the K_V, EAG, and KCNQ families. Of these, only members of the K_V family have consistently been found to be expressed in vascular SMC. So far, only four subfamilies of functional α subunit proteins are expressed in vascular smooth muscle cells, Kv1–Kv4. However each of these α subunit subfamilies is composed of multiple members, due to alternative splicing [35]. Other important factor that contributes to the large channel diversity is the presence of cytosolic accessory beta subunits, which are associated with each alpha-subunit [11, 27]. Apparently, the K_V channels need the ancillary subunits to fulfill different physiological functions. This fact could be due to the large structural diversity observed with auxiliary subunit structures, which range from proteins with transmembrane segments and extracellular domains to purely cytoplasmic proteins. These subunits modulate K_V channel gating and can also have a great impact on channel assembly, on channel trafficking and on targeting K_V channels to different cellular compartments [36].

Some studies have reported a differential expression of K_V channel genes (mRNA and protein) at the vascular level. Xu et al. (2000) observed in SMC from rat tail artery the mRNA expression of Kv1.2, Kv1.3, Kv1.4, Kv1.5, Kv2.1, Kv2.2,Kv3.2, Kv3.3, Kv3.4, Kv4.1- Kv4.3, Kvbeta1, Kvbeta2, and Kvbeta3, but only observed the presence of Kv1.2, Kv1.3, Kv1.5, and Kv2.1 proteins[37], and this could be linked to the large diversity and complexity of these channels.[12, 37, 38]. Furthermore, heterotetrameric and homotetrameric association of K_V channel subunits also contributes to a vast array of K_V current types. For example, a channel with two Kv2.1 and two Kv1.5 subunits may functionally differ from a tetramer channel with three Kv2.1 subunits and one Kv1.5 subunit. In this sense, the K_V currents dramatically differ in terms of kinetics, amplitude and response to drugs. This is possible the cause of the differences among species and vessels in the channel kinetic parameters, such as the voltage dependence for activation and inactivation, the single-channel conductance and the sensitivity to inhibitors [12].

Pharmacology of K_V Channels

An interesting issue in K_V channel subtypes is their sensitivity to pharmacological modulators. The compound 4-aminopyridine (4-AP) has been used in many studies with vascular smooth muscle as a K_V channel blocker, in order to separate the K_V current from BK_{Ca} current, which is also activated by membrane depolarization [1, 17]. Moreover, electrophysiological recordings have revealed that channels constituted by subunits encoded by Kv1.2 and K_V1.5 genes are relatively sensitive to 4-AP [39], whereas that with Kv2 subunits are inhibited more effectively by TEA [40]. On the other hand, channels with Kv1.2 subunits are quite sensitive to charybdotoxin [41], Kv1.3, and Kv1.6 [42], commonly used to inhibit BK_{Ca} channels (Table 1). However, the compound 4-AP is quite specific when used to the concentrations between 0.3 to 1.1mM, and has no effect on BK_{Ca} or Kir channels at these concentrations [11]. Other inhibitors are used for other K^+ channels, including Ba^{2+} (50μM for Kir channel), glibenclamide (10 μM for K_{ATP} channel) and iberiotoxin (100nM for BK_{Ca} channel), but at these concentrations none of them affect K_V current activity [11, 43]. In summary, the differences in sensitivity among the K_V channels might be attributed to the differential expression of different subunits that constitute distinct K_V channel subtypes in animal species [12].

Regulation of K_V Channels

Several physiologic effects seems to be mediated in part through the K_V channels function. The K_V channel phosphorylation by kinases and phosphatases has been shown to be very important physiologically [44, 45], and multiple sites for phosphorilation by several kinases have been identified [35] (Figure 1).

Table 1. Kv channels inhibitors

Kv1.1	4-AP[22]; TBA[22]; DTX[22]; Veramapil[22]; TEA [41]; Diltiazem [41]; Nifedipine [41]; Felcainide [41]; Quinine [35]; CTX [41]; DTx [41]; kaliotoxin [41]; MCDP [41]; Maurotoxin [35]; Noxiustoxin [41]; Hongotoxin[172]; Margatoxin[172]; Agitoxin [173]; ViTx[174]; M-2,5-D [175]; CC-5 [176]; 1,3-D-2CC-2 [176]; N-T2[22].
Kv1.2	4-AP[22]; TBA[22]; DTX[22]; Veramapil[22]; TEA [41]; Diltiazem [41]; Nifedipine [41]; Felcainide [41]; CTX [41]; DTx [41]; kaliotoxin [41]; MCDP [41]; Maurotoxin [173]; Noxiustoxin [41]; Hongotoxin[172]; Margatoxin[172]; Pi4 [177].
Kv1.3	4-AP[22]; TBA[22]; DTX [22]; Veramapil[22]; TEA [41]; Diltiazem [35]; Nifedipine [41]; Felcainide [41]; Quinine [35]; CTX [41]; DTx [41]; kaliotoxin [41]; MCDP [41]; Maurotoxin [173]; Noxiustoxin [41]; Hongotoxin[172]; Margatoxin[42]; Agitoxin [173]; stichodactlyatoxin[178]; ViTx[174]; CP-339818 [22]; UK-78282 [22]; Correolide [22]; PAP-1 [179]; khellinone chalcone [180] ; 4-SK [181] ; clofazimin [182].
Kv1.4	4-AP[22]; TBA[22]; DTX [22]; Veramapil[22]; TEA [41]; Felcainide[183]; Quinine [35]; DTx [34]; MCDP [35].
Kv1.5	4-AP[22]; TBA[22]; DTX [22]; Veramapil[22]; TEA [41]; Diltiazem [41]; Nifedipine [41]; Felcainide [41]; Quinine [39]; CTX [41]; DTx [41]; kaliotoxin [41]; MCDP [41]; Noxiustoxin [41]; Arylsulphonamidoindane[184]; AVE0118[185]; AVE1231[186]; Vernakalant[187]; ISQ-1[188]; TAEA[188]; tetrazole derivative[189]; DPO-1[190]; XEN-D0101[191].
Kv1.6	4-AP[22]; TBA[22]; DTX [22]; Veramapil[22]; TEA[192] ; CTX [34]; DTx [34]; MCDP [35]; Noxiustoxin [35]; Hongotoxin[172]; Margatoxin [35].
Kv1.7	4-AP[22]; TBA[22]; DTX [22]; Veramapil[22]; TEA [34] ; Margatoxin [34]; ShKTx [34]; Noxiustoxin [34].
Kv1.8	4-AP[22]; TBA[22]; DTX [22]; Veramapil[22].
Kv2.1	4-AP[22]; TBA[22]; DTX [22]; Veramapil[22]; TEA [35]; Hanatoxin [34]; ScTx[193].
Kv2.2	4-AP[22]; TBA[22]; DTX [22]; Veramapil[22]; TEA [34]; Quinine [194]; Hanatoxin [34]; ScTx[193].

Table 1. (Continued)

Kv3.1b	4-AP[22]; TBA[22]; DTX [22]; Veramapil[22]; TEA [41]; Diltiazem [41]; Nifedipine [41]; Felcainide [41]; Quinine [35]; CTX [41]; DTx [41]; kaliotoxin [41]; MCDP [41]; Noxiustoxin [41]; Chromakalin [34].
Kv3.2	4-AP[22]; TBA[22]; DTX [22]; Veramapil[22]; TEA [34]; Quinine [35].
Kv3.3	4-AP[22]; TBA[22]; DTX [22]; Veramapil[22]; TEA [34].
Kv3.4	4-AP[22]; TBA[22]; DTX [22]; Veramapil[22]; TEA [34]; Quinine [35]; BDS[195].
Kv4.1	4-AP[22]; TBA[22]; DTX [22]; Veramapil[22]; HmTx[193]; HpTX[34].
Kv4.2	4-AP[22]; TBA[22]; DTX [22]; Veramapil[22]; Felcainide[183]; Quinine [183]; Hanatoxin [34]; ScTx[193]; phixotoxin [34]; ShKTx [41].
Kv4.3	4-AP[22]; TBA[22]; DTX [22]; Veramapil[22]; ShKTx[41].

4-AP (4-aminopyridine); TBA (tetrabutyl ammonium); CTX (Charybdotoxin); MCDP – mast cell degranulating peptide; Pi4 – pandinus imperator toxin; ViTx- conis virgo toxin; M-2,5-D (Methyl 2,5-dihydroxycinnamate); 4-SK4 (substituted khellinone); CC-5 (cylohexadione compound-5); 1,3-D-2CC-2 (1,3-dione-2-carboxamide compound-2); N-T2 (N-tosyl-2-(3-tosylureido)-7,8-dihydro-1,6-naphthyridine-6(5H)-carboxamide compound-6); ShKTx (stichodactlyatoxin); DTx (α-dendrotoxin); DTX (d-tubocurarine), ScTx- stromatoplema calcata toxin; BDS – blood depressing substance; HmTx – heteroscodra maculata toxin; HpTX (heteropodatoxin); ShKTx (stichodactlyatoxin).

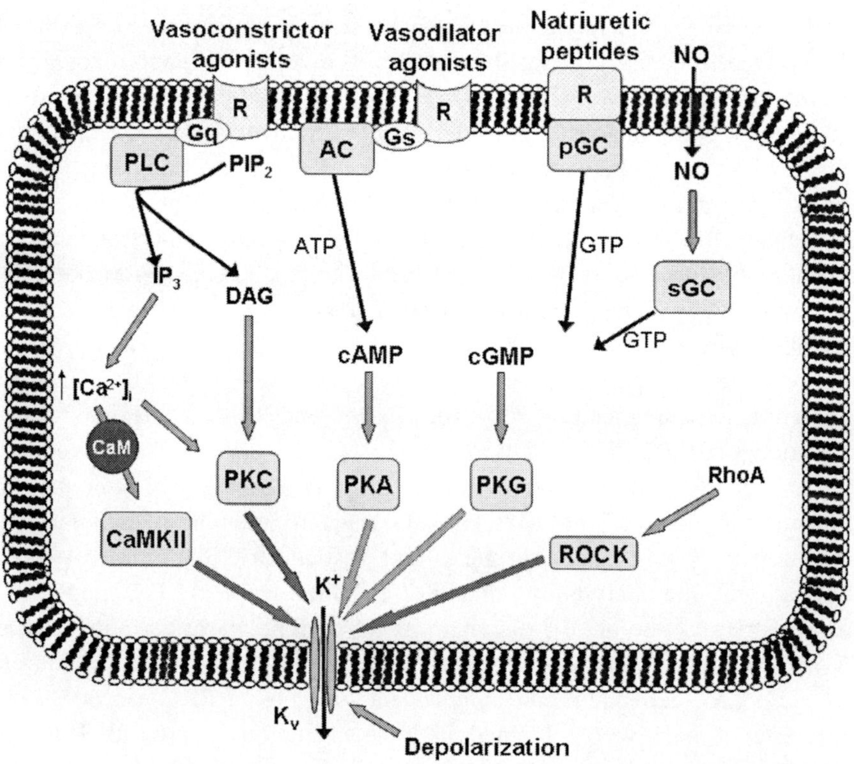

Figure 1. Mechanisms for K_V channels modulation in vascular smooth muscle cells. The K_V channels is regulated by vasoconstrictors, vasodilators and by depolarization. Green arrows means stimulation and red arrows means inhibition. AC, adenylate cyclase; CaM, calmodulin; pGC, particulate guanylate cyclase; sGC, soluble guanylate cyclase; NO, nitric oxide; DAG, diacylglycerol ; IP_3, inositol 1,4,5-triphosphate; R, receptor; G, G-protein; cAMP, cyclic adenosine 3',5'-monophosphate; cGMP, cyclic guanosine 3',5'-monophosphate; ATP, adenosine 5'-triphosphate; GTP, guanosine 5'-triphosphate; PKA, cAMP-dependent protein kinase; PKG, cGMP-dependent protein kinase; PKC, protein kinase C; CaMK II, Ca^{2+}-calmodulin-dependent protein kinase II; ROCK, Rho-associated protein kinase.

These channels could be involved in the regulation of artery contraction by drugs, such as bosentan, a competitive and non-selective endothelin receptors (ET_A and ET_B) inhibitor. Thus, the vasodilators activate directly or indirectly these channels, while vasoconstrictors close these channels [5, 19, 46]. Most studies agree that some vasoconstrictors induce K_V channel inhibition through the activation of phospholipase C (PLC) and protein kinase C (PKC). Some membrane receptors are coupled, through a GTP-binding protein (Gq), to phospholipases. The activation of these receptors generate the second messengers diacylglycerol and inositol 1,4,5-triphosphate (IP_3). The IP_3 produces increase on citosolic Ca^{2+} levels and this Ca^{2+} and diacylglycerol activate protein kinase C (PKC) [47-50]. Several PKC isoforms, with different Ca^{2+}-dependence, such as α, β, ε and ζ have been identified in vascular smooth muscle [51, 52]. To become activated, some PKC require Ca^{2+}, diacylglycerol and also phosphatidylserine and these are designated as classic PKCs (α and β). Other PKC (ε), designated as novel, require diacylglycerol and phosphatidylserine, but not Ca^{2+}. Also, the atypical PKC (ζ) do not require Ca^{2+} or diacylglycerol, but are instead activated by phosphatidylserine and phosphatidyinositol 3,4,5-triphosphate (PIP_3) [11, 53]. For example, in rat pulmonary arteries, the activation of 5-HT_{2A} receptors inhibits the K_V currents concomitant with membrane depolarization and this is mediated by the activation of PLC, PKC and tyrosine kinases, and the PKC involved is the classic Ca^{2+}-dependent protein kinase [54]. Differently, thromboxane A2-induced inhibition of the K_V channels leads to membrane depolarization, activation of L-type Ca^{2+} channels, and vasoconstriction of rat pulmonary arteries via the activation of PKC(ζ), the atypical PKC [55]. The vasoconstrictor endothelin-1 also inhibits K_V currents in rat pulmonary arterial SMC. This inhibition requires the activation of PKC secondary to activation of PLC, and Ca^{2+}-dependent and independent subtypes of PKC can be involved [56]. Similar data were obtained in human pulmonary arterial SMC [57]. Several authors also show that Angiotensin II also inhibits K_V currents in rabbit portal vein SMC [58] and in mesenteric arterial SMC [59], and this inhibitory effect was completely dependent on the activation of Ca^{2+}-independent PKC subtypes. Structural studies demonstrated that consensus PKC phosphorylation sites are present in almost all K_V alpha subunit sequences, Besides, it has been also demonstrated that the modulatory effect of β1.3 on Kvα1.5 is dependent on the phosphorylation of the β subunit by PKC [60]. After performing yeast two-hybrid system studies, some authors suggested that two related proteins, ZIP1 and ZIP2, are to required to couple PKC with the K_V channel α/β complex [35, 61]. These results indicate that the channel regulation involves multiple proteins. Thus, more studies are needed to better clarify the molecular and structural mechanism that modulated the K_V channels by this kinase.

Concerning the effects of vasodilators on K_V channels, the main kinases involved are cAMP- and cGMP-dependent protein kinases, PKA and PKG respectively. The activation of PKA with a membrane-permeable cyclic adenosine monophosphate analog or the dialysis of the catalytic PKA subunit into cells, dramatically increase K_V current [35, 60, 62]. The β-adrenergic receptors activation with isoproterenol or adenylyl cyclase activation with forskolin increases K_V currents in rabbit portal vein SMC [8, 63, 64]. Some authors performed studies with recombinant proteins and have demonstrated that PKA phosphorylates only the α subunit of K_V channels [65]. Other authors suggested that the phosphorylation of α subunit is achieved only in the presence of an accessory subunit [66, 67] or after the phosphorylation of an accessory subunit from the heteromeric complex [35, 60].

Concerning PKG, few studies were performed and suggested that the role of this kinase is similar to the observed for PKA. Nitric oxide (NO) and sodium nitroprusside (SNP) stimulate K_V currents in cultured pulmonary artery SMC, leading to hyperpolarizantion and inhibition of spontaneous Ca^{2+} dependent action potentials [68]. Nitric oxide and SNP also relax pulmonary artery rings pre-contracted with KCl, and this effect was inhibited by 4-AP and charybdotoxin, suggesting the participation of both K_V and BK_{Ca} channels in this outcome [69]. Sobey and Faraci (1999) demonstrated that SNP, acetylcholine and 8-bromo cyclic guanosine 5′-monophosphate (GMP) induce rat basilar artery relaxation, an effects that was inhibited by pre-treatment with the K_V channel inhibitor 4-AP [70]. Tanaka et al. (2006) also demonstrate the participation of K_V channels in the relaxation of rat aorta by NO and by atrial natriuretic peptide (ANP) [71]. More recently, Cairrao et al. (2010) demonstrated that in human umbilical artery SMC ANP, but not SNP, stimulate the activity of BK_{Ca} and K_V channels due to a PKG action [17]. Furthermore, other known vasodilators, such as prostacyclin and adenosine, activate the K_V currents in rabbit cerebral and coronary arterioles, respectively [72, 73]

Concerning phosphorylation, Kv1.5 and the Kvβ1 subfamilies, that are present in vascular SMC, have consensus PKG phosphorylation sites that may be responsible for the channel activation [35].

Other kinases could also phosphorylate K_V channels and change their functional properties, like for example Rho kinase [74], tyrosine kinases [75, 76] and calcium-calmodulin-dependent protein kinase II (CaMKII) [76-79]. Relatively to the Rho kinase, Luykenaar et al. (2004) observed the role of this kinases in the vasoconstriction of cerebral arteries mediated by uridine 5’-triphosphate (UTP).

This effect is mediated via P2Y receptors through the activation of voltage-gated Ca^{2+} channel and was potently inhibited by Y-27632, a Rho-kinase inhibitor. Furthermore, UTP decreased K_V currents in myocytes from theses arteries an this was inhibited by Y-27632 and by the RhoA inhibitor C3 exoenzyme [74]. Interestingly, tyrosine kinases activated by G protein-coupled receptors suppress Kv1.2 channel activity, an effect mediated also by RhoA [75].

Other studies showed that K_V channels from rat pituitary cell line are inhibited by thyrotropin-releasing hormone through a process blocked by RhoA inhibitors [80].On the other hand, CaMKII, is emerging as an important regulatory mechanism of delayed rectifier K channels in many cell systems. Some studies in a variety of cells suggest that activation of CaMKII slows inactivation gating of delayed rectifier channels and increases the sustained component of these currents in native cells and in heterologous expression systems [77-79].

Recent studies have shown that inhibition of CaMKII accelerates the inactivation kinetics of transient outward current in human atrial myocytes, and this kinase regulates Kv4.3 kinetics by direct phosphorylation of the α-subunits at Ser550 [76].

The knowledge about this channel in the vascular SMC is very important because several vascular diseases, including hypertension, atherosclerosis, diabetes and chronic pulmonary hypoxia have modification/alteration in these channels [10, 35].

In most cases, K_V currents are functional down-regulated, as in in aorta and renal, tail and mesenteric arteries of hypertensive rat, and in pulmonary artery of human and rat with pulmonary hypertensive disease [35]. Targeting processes that increase the translation or plasma membrane transport of K_V channels could increase the functional expression of native K_V currents. Expression of accessory subunits that increases total expression of alpha subunits without increasing K_V currents inactivation would be another means of functionally up-regulating K_V currents. Expression of an accessory subunit that change the voltage dependence of K_V channels activation in the negative voltage direction would theoretically produce membrane potential hyperpolarization and inhibition of vascular SMC force maintenance [35].

3. Calcium-Activated K^+ Channels (K_{ca})

In general, it is well established that K_{Ca} channels are important effectors in the control of vascular tone and blood pressure by mediating membrane hyperpolarization in response to elevations of intracellular Ca^{2+} and thereby counteracting smooth muscle contractility. Three broad categories of K_{Ca} channels have been identified and classified according to their conductance: large-conducatance K_{Ca} channels, 100-300pS (BK_{Ca}) [81], intermediate-conductance K_{Ca} channels, 25-100pS (IK_{Ca}) [82], and small-condutance K_{Ca} channels, 2-25pS (SK_{Ca}) [83, 84]. The SK_{Ca} and IK_{Ca} channels are predominately expressed in the endothelial cells, and IK_{Ca} were also expressed in the proliferating vascular SMC.

In the vascular SMC the dominant K_{Ca} are the BK_{Ca} channels, which are also known as BK, Slo and MaxiK channels. These channels have been extensively studied because their primordial contribution in the control of vascular tone [85, 86]. They are activated in response to membrane depolarization and by binding of intracellular Ca^{2+} and Mg^{2+}, and may also contribute to the modulation of the membrane potential in small myogenic vessels [1, 5, 10, 11, 87].

The activation of BK_{Ca} channels hyperpolarizes the membrane and promotes closing of voltage dependent Ca^{2+} channel, thereby opposing the role of these Ca^{2+} channels in vasoconstriction [86, 88]. Several studies showed that the BK_{Ca} channels are activated by highly localized Ca^{2+} sparks produced by transient Ca^{2+}-release events from the sarcoplasmic reticulum and/or Ca^{2+}-influx through voltage dependent calcium channels. The resulting spontaneous transient outward currents (STOC) produce hyperpolarization/repolarization and counteract thereby depolarization-activated voltage dependent calcium channels activity.

As a consequence of voltage dependent calcium channel inactivation, the intracellular concentration of Ca^{2+} decreases resulting in the inactivation of the myosin light-chain kinase and thus in relaxation [43, 88-90]. The BK_{Ca} may also contribute to the alterations in the vascular tone in pathophysiological states, such as hypertension, stroke, atherosclerosis, diabetes and complications of cardiovascular surgery [91]. Thus, BK_{Ca} were suggested as possible therapeutic targets for the treatment of some cardiovascular diseases [86].

Structure of Bk_{ca} Channels

Similarly that K_V channels, BK_{Ca} channels have a pore-forming α-subunit and a regulatory β-subunit [11, 84, 92]. The α-subunits contain six transmembrane-spanning domains (S1–S6), including a voltage sensing (S1-S4; designated as VSD) and pore-gate (S5-S6, designated as PGD) domains [93]. However, the α-subunits, which are produced from a single gene (slo) by alternative splicing [8, 11, 94], contain an additional seventh transmembrane region (S0) that confers the N-terminus to the extracellular side and a large cytoplasmic domain containing ~800 amino acids [93]. The sequence and predicted secondary structure of the cytoplasmic domain are homologous to the regulatory domain for K^+ conductance (RCK domain) that is found in a number of K^+ channels and transporters [93, 95]. This cytosolic domain comprises two RCK (regulator of K^+ conductance) domains, RCK1 and RCK2 [96]. These domains contain two putative high affinity Ca^{2+} binding sites: one in the RCK1 domain at position Asp362/Asp367 (this numbering scheme is based on the mbr5 sequence of mouse α subunit) and the other in a region termed the Ca^{2+} bowl that contains a series of Asp residues, located in the RCK2 domain [87, 96]. In addition, recent studies have showed that the Mg binding site is placed in the interface between the VSD (Asp99 and Asn172) and the cytosolic domain (Glu374 and Glu399) [87]. However, several questions remain unanswered, like e.g. how the voltage and metal ion binding open the activation gate. In conclusion, a functional BK_{Ca} channel can be divided into three major structural domains: the voltage-sensor domains (VSD), the C-terminal cytoplasmic domain senses various intracellular ligands, and the pore gate domain (PGD) that controls ion permeation in response to different stimuli [93]. In addition, there are four β-subunit isoforms (β1-4), each with two transmembrane domains, which may be associated to the α-subunits in a 1:1 ratio [87, 91, 97]. Among the four β isoforms, the β1 subunit is the predominant in vascular smooth muscle [11, 87, 98, 99]. The major function of the β-subunits is to enhance the Ca^{2+} sensitivity of the channel, and thus increase STOC activity and repolarizations and thereby facilitate relaxations [88].

Pharmacology of Bk_{ca} Channels

Like for K_V channels, BK_{Ca} channel are sensitible to specific pharmacological modulators. The blockers of these channels are pharmacological tools used to examining the structural characteristics of the BK_{Ca} channels and to understand their role in physiological process [84]. The BK_{Ca} blockers are tetraethylammonium (TEA; <1 mM), charybdotoxin, iberiotoxin, penitrem A, and paxilline, but only the last three compounds are selective for the BK_{Ca} channels [84]. On the other hand, these channels can be activated by agents that are called BK_{Ca} openers or BK_{Ca} activators. The BK_{Ca} openers increase the efflux of potassium ions leading to hyperpolarization, and thus decrease the cell excitability and/or smooth muscle relaxation. Thus, the activation of these channels has potential therapeutic implications because of their involvement in several pathologies, such as hypertension. The BK_{Ca} openers comprise a large series of synthetic benzimidalone derivatives, such as NS1619 and NS004, the biaryl amines, such as mefenamic and flufenamic acids, the biarylureas, such as NS1609, the pyridyl amines, natural modulators such as resveratrol and (dedydrosoyasaponin-1, DHS-1), and flavonoides [84]. BK_{Ca} channels are usually identified by their pharmacological profile, e.g., insensitivity to apamin (selective SK_{Ca} channel blocker; [100]) in combination with high sensitivity to iberiotoxin, which does not block SK_{Ca} or IK_{Ca}. These channels are not blocked by glibenclamide, a K_{ATP} channels blocker, or by low concentrations (100μM) of external Ba^{2+}, which inhibit K_{ATP} and K_{IR} channles [1]. Table 2 summarizes the tools used to study the function of BK_{Ca} channels.

Regulation of Bk_{ca} Channels

The BK_{Ca} channels regulate membrane voltage and intracellular Ca^{2+} concentration. The function of these channels in vascular SMC is regulated by a diversity of vasodilators and vasoconstrictors. Along with other ion channels, kinases, phosphatases, and regulatory proteins, these channels seem to be organized in macromolecular complexes that allow rapid and specific activation by distinct signal transduction pathways [86] (Figure 2). Smooth muscle vasodilators such as β_2-adrenergic agonists, nitric oxide, prostaglandin I_2, arachidonic acid, calcitonin gene-related peptide, and carbon monoxide, activate BK_{Ca} channels [1, 101]. Moreover, potent vasoconstrictors like angiotensin II, 5-HT, phenylephrine and thromboxane A_2 can inhibit BK_{Ca} activity [91, 101, 102]. In general, most of the agonistic actions involve intracellular signalling pathways like: (i) Ca^{2+} sparks that appear to be involved in the action of vasodilators via a negative feedback mechanism; (ii) kinases that phosphorylate the channel protein and (iii) G proteins that regulate channel activity independent of downstream kinases [101].

The BK_{Ca} channels are also modulated by a endogenous molecules including steroids, fatty acids, phosphatidylinositol 4,5-bisphosphate (PIP_2) and ROS [91, 103]. For steroid hormones, modulation of BK_{Ca} activity has been shown to occur through genomic [104-106] and non-genomic mechanisms [17, 107, 108]. Concerning the latter, estrogen-induced relaxation of coronary arteries activated BK_{Ca} via cGMP [108] and this is also the case for testosterone-induced relaxation of human umbilical artery [17].

Table 2. Drugs modulating BK_{Ca} channels

Blockers	Effects (concentration for supramaximal effect)	References
Iberotoxin	Specific BK_{Ca} channel blocker in SMC and ECs; α-subunit selective (<10nM)	[11, 86, 196-198]
Paxilline	Specific BK_{Ca} channel blocker in SMC and ECs; α-subunit selective (300nM)	[86, 199-201]
Penitrem	Specific BK_{Ca} channel blocker in SMC; not studied in ECs (100nM)	[86, 201-203]
Charybdotoxin	Blocks BK_{Ca},IK_{Ca} and SK_{Ca} channels in SMC and ECs; α-subunit selective (100nM)	[86, 204-208]
TEA	Partial block of BK Ca in SMC and ECs; non-selective block of potassium channels; (100mM)	[86, 206, 209-211]
BmBKTx1	Selectively inhibit α-subunit (400nM)	[84, 212]
BmP09	Selectively blocks BK_{Ca} currents (200nM)	[84, 213]
Slotoxin	Selectively inhibit α-subunit and can distinguish between the α, α+β1 and α+β4 complexes (100nM)	[84, 214]
Openers	Effects (concentration for supramaximal effect)	References
NS004	Specific BK_{Ca} channel opener in SMC; blocks Kv and K_{ATP} (at 50 μM); not studied in ECs; α-subunit selective (50μM)	[84, 215-217]
NS1608	Specific BK_{Ca} channel opener in SMC; not studied in ECs; α-subunit selective (10μM)	[84, 216]
NS1619	Specific BK_{Ca} channel opener in SMC and in ECs; α-subunit selective (30μM)	[203, 211, 218]
DHS-I	Specific BK_{Ca} channel opener in SMC; ineffective in ECs β-subunit selective (250μM)	[219, 220]
Resveratrol	IK_{Ca} and BK_{Ca} channel opener in SMC and ECs; Inhibition of Na^+ currents (50μM)	[200, 211, 221, 222]
Mefenamic acids	BK_{Ca} opener in a variety of cells; block K_V, K_{ATP}, HERG, calcium-activated chloride channels and non-selective cation channels (75μm to 1mM)	[84, 223-226]
Flufenamic acids	BK_{Ca} opener in a variety of cells; block K_V, K_{ATP}) (75μm to 1mM)	[84, 223-227]
Niflumic	BK_{Ca} opener in a variety of cells; block K_V, K_{ATP}, HERG, calcium-activated chloride channels and non-selective cation channels (75μm to 1mM)	[84, 223-227]

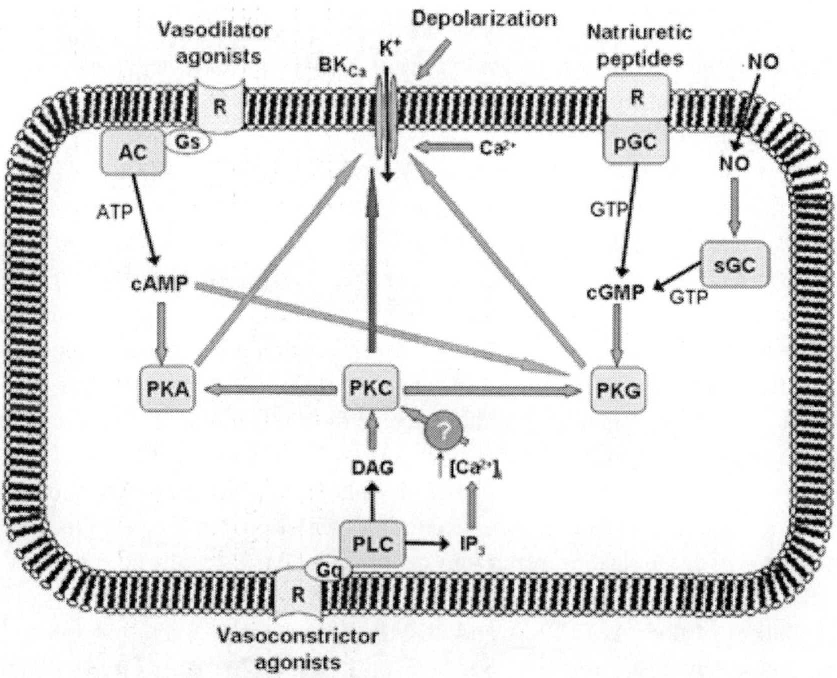

Figure 2. Mechanisms for BK_{Ca} channels modulation in vascular smooth muscle cells.. The BK_{Ca} channels is regulated by vasoconstrictors, vasodilators, depolarization and calcium. Green arrows means stimulation and red arrows means inhibition. AC, adenylate cyclase; pGC, particulate guanylate cyclase; sGC, soluble guanylate cyclase; NO, nitric oxide; DAG, diacylglycerol ; IP_3, inositol 1,4,5-triphosphate; R, receptor; G, G-protein; cAMP, cyclic adenosine 3',5'-monophosphate; cGMP, cyclic guanosine 3',5'-monophosphate; ATP, adenosine 5'-triphosphate; GTP, guanosine 5'-triphosphate; PKA, cAMP-dependent protein kinase; PKG, cGMP-dependent protein kinase; PKC, protein kinase C.

In cerebral arteries it is likely that NO increases cGMP and this stimulated BK_{Ca} channels[109], although NO has been suggested to exert direct effects on these channels [110].

Functionally, BK_{Ca} beta subunits are reported to influence the kinetics and calcium sensitivity of channel gating, the responses of BK_{Ca} activity to pharmacologic modulators, and the trafficking of channels to the plasma membrane [91]. Recent studies showed that the opening of BK_{Ca} may be due to an action on the β1 subunit and these effects were demonstrated using 17β-estradiol and tamoxifen in murine colonic myocytes [103] and in dog SMC [111], respectively. Moreover, diffeent subunit stoichiometries (i.e. alpha:beta subunit ratio) may exist, and a full complement of four beta1 subunits are required for maximal effect. Genetic deletion of the beta1 subunit in mice leads to hypertension and increased contractility of VSM, whereas a gain-of-function mutation in the beta1 subunit is associated with lower blood pressure in humans [91].

Some kinases can modulate the BK_{Ca} activity. In general the endogenous vasodilators mediate the increase of cAMP and /or cGMP, and vasoconstrictors induce the PKC activation or the Src tyrosine kinase activation [101, 102], although the relevance of this last pathway has been controversial [91]. Thus, in most cases, PKA and PKG activate the BK_{Ca} channel by increasing its open probability, thereby shifting its voltage-activation curve toward more negative membrane potentials but without affecting the single-channel conductance or the voltage sensitivity, represented by the slope of the voltage-activation curve [92, 112, 113]. PKA or PKG can also indirectly activate BK_{Ca} channels by phosphorylation of phospholamban, which increases the sarcoplasmic reticulum calcium load by eliminating the inhibition of sarcoplasmic reticulum calcium ATPase, resulting in an elevated calcium spark frequency and amplitude [92, 114-116] In addition, the stimulation of the cAMP/PKA and cGMP/PKG cascade likely phosphorylates BK_{Ca} channel α-subunits at Ser-869 and Ser-1072, respectively, thereby activating the channel. On the other hand, PKC phosphorylation may inhibit the BK_{Ca} channel in smooth muscle [117-119]. It has also been reported that phosphorylation by PKC determines the susceptibility of the channel to be stimulated by either PKG or PKA [120]. This is consistent with the finding that PKC activates BK_{Ca} channels via PKG in rat pulmonary arterial myocytes [92, 119]. In summary, the current knowledge about enzymatic regulation of BK_{Ca} suggests that both PKG and PKA activate BK_{Ca} channels, whereas the role of PKC and Src tyrosine kinase need to be further investigated.

4. Inward Rectifier K^+ Channels (Kir)

The main functional characteristic of these channels is that K^+ flow can enter the cell, in the opposite direction than for the other K^+ channels. For K^+ channels, inward rectification means that at any given driving force (voltage), the inward flow of K^+ ions is greater than the outward flow. The functional role of Kir channels depends critically on their degree of rectification [121, 122]. The activation of these channels is not dependent on membrane potential per se, and the channels are not voltage-gated. Rather, activation depends on the difference between the membrane potential and the equilibrium potential for K^+ [123]. Recent evidences strongly suggest that the physiological role of Kir channels in vascular smooth muscle involves maintenance of the resting membrane potential and resting tone in small diameter coronary and cerebral arteries [124]. Kir channels were first described in vascular SMC approximately 20 years. They were described as present only in certain small-diameter cerebral and submucosal arterial and in coronary arterial smooth muscle [124].

Structure of Kir Channels

These channels have four subunits forming two transmembrane spanning helices (TM), called TM1 and TM2, that are connected to the pore region containing the selectivity filter for K^+ ions [25, 125]. TM1 (the outer helix) makes contact with TM2 (the inner helix), which lines the pore of the channel and forms an "inverted tepee" in a closed channel conformation [123].

These channels are encoded by members of the Kir gene family. The Kir2.0 subfamily encodes the classical inward rectifiers found in several types of cells, including the vascular SMC [8, 11, 126]. Four isoforms of this family have been identified, Kir2.1 to Kir2.4, however in vascular SMC only the Kir2.1, 2.2 and 2.4 have been identified [11]. The Kir2.2 was only detected in cerebral arteries [127] and the Kir2.4 expression has been only detected in SMC of human pulmonary artery [128]. Additional studies are required to characterize the expression and function of each Kir subtypes.

Pharmacology of Kir Channels

Some ions are able to block the Kir channels and are usually used as pharmacological tools to study the physiological role of these channels. Some of the blockers are ions such as Ba^{2+}, Cs^+, Mg^{2+}, and Na^+ that were used in patch clamp [122, 123], although Ba^{2+} is the most used inhibitor. In coronary and cerebral arteries the half-inhibition values are 2.1 μM and 2.2 μM at –60 mV, respectively. [11, 126, 129, 130]. In fact, only at concentrations below 50 μM, Ba^{2+} is a selective blocker of Kir channels in vascular smooth muscle. These ion is also an effective blocker of K_{ATP} channels (K_d, 200 μM at –60 mV), BK_{Ca} channels (K_d>10mM), and Kv channels (K_dC>1 mM) [1, 11, 129]. Interestingly, external Ca^{2+} and Mg^{2+} also blocked Kir channels in rat coronary arterial smooth muscle [124, 129]. Both Ca^{2+} and Mg^{2+} also blocked the whole-cell inward current in cardiac muscle with very little voltage dependence, suggesting that the Kir channels in cardiac myocytes and coronary arterial cells exhibit a similar type of low-affinity inhibition by Ca^{2+} and Mg^{2+} [124, 125, 129]. Furthermore, external Cs^{+} blocked Kir currents, with a K_d of 2.9 mM at −60 mV [11]. The inhibition of Kir currents in vascular SMC is highly voltage dependent with rapid kinetics [124, 129, 131]. A similar level of voltage dependence has been reported for the blockage of inward rectification by Cs^{+} by produces a voltage-dependent block of the delayed, outward K currents, though the ratio of Cs^{+} to K^{+}, in starfish oocytes, skeletal muscle and cloned IRK1 channels. This suggests that multiple ions can simultaneously occupy the channels [8].

Regulation of Kir Channels

The effects of vasoconstrictors and vasodilators on Kir channels were not studied as much as for the other K^{+} channel types. In general, the lack of information on Kir channel regulation makes difficult to determine the exact physiological and pharmacological properties of these channels [124].

In relation to the effect of the vasoconstrictors (Figure 3) on Kir channel in vascular SMC, only a few studies have focused this matter [127, 132-134]. Some studies demonstrate that ET-1 and Ang II can inhibit Kir channels and induce vasoconstrictors [132, 133]. Electophysiological studies in vascular myocytes o revealed that these two vasoconstrictors modulate various types of ion channels, including K_{ATP}, BK_{Ca}, K_V, Ca^{2+} channels, Cl^{-} channels and nonselective cation channels[124, 135]. The regulation of these channels by ET-1 and Ang II is closely related to PKC activation. Recent data obtained working with small-diameter arteries suggested that the inhibitory effects of Ang II and ET-1 on Kir channel function is closely associated with Ca^{2+}-dependent PKCα activation [132, 133], because the inhibitory effect of Ang II and ET-1 was reduced by intracellular and extracellular Ca^{2+} free condition, which inhibits Ca^{2+}-dependent PKC isoforms alpha and beta [124]. Moreover, the inhibitory effect of Ang II was not affected by a selective inhibitor of PKCε. The mechanism for regulation of of Kir channels by PKCα in small-diameter arteries is unclear, although the dominant isoforms of PKC are likely to be involved.

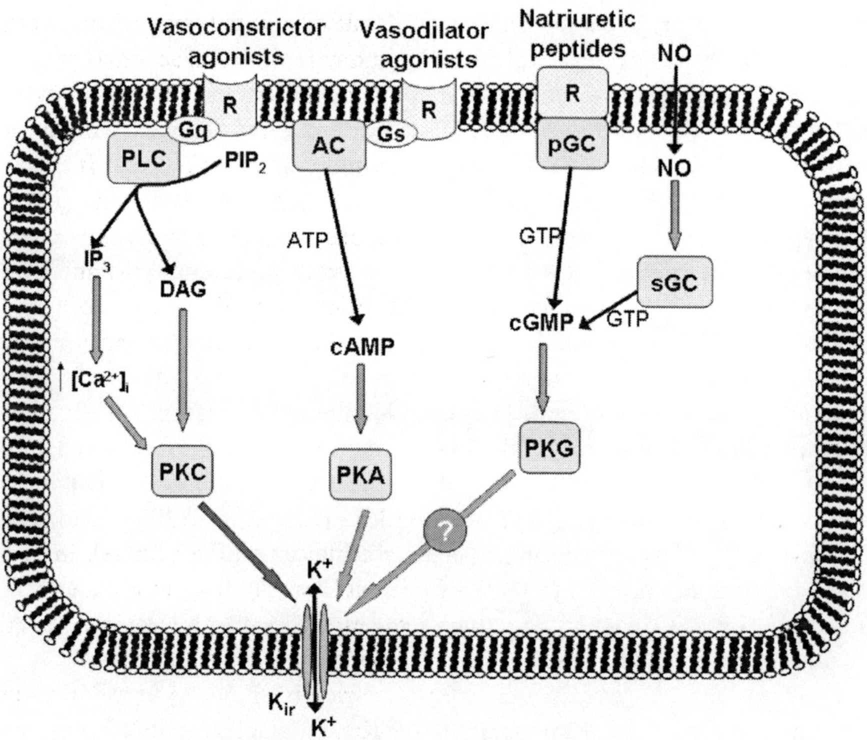

Figure 3. Mechanisms for Kir channels modulation in vascular smooth muscle cells.. The Kir channels is regulated by vasoconstrictors and vasodilators. Green arrows means stimulation and red arrows means inhibition. AC, adenylate cyclase; pGC, particulate guanylate cyclase; sGC, soluble guanylate cyclase; NO, nitric oxide; DAG, diacylglycerol ; IP_3, inositol 1,4,5-triphosphate; R, receptor; G, G-protein; cAMP, cyclic adenosine 3',5'-monophosphate; cGMP, cyclic guanosine 3',5'-monophosphate; ATP, adenosine 5'-triphosphate; GTP, guanosine 5'-triphosphate; PKA, cAMP-dependent protein kinase; PKG, cGMP-dependent protein kinase; PKC, protein kinase C.

Consistent with this idea, it was been suggested that large-diameter arteries exhibit Ca^{2+}-independent and PKC-dependent contractions due to their relatively higher PKCε levels and relatively lower PKCα levels [11].

Recently, it was shown that Kir channel activity is unaffected by the vasoconstrictors uridine triphosphate (UTP) and thromboxane A2 agonist (U46619) in cerebral smooth muscle [127], which have been shown previously to strongly depolarize and constrict intact cerebral arteries. Both vasoconstrictors seems to activate PKC in resistance arteries; however, they do not appear to affect PKC in the cerebral circulation. In these resistance arteries the contraction mediated by UTP and U46619 seems to be induced by RhoA and Rho kinase pathway [124].

Recent studies demonstrated that some vasodilators including bradykinin, adenosine, and cicaprost, exhibited local and remote vasodilatation through activation of Kir channels in vascular arterioles [124, 130, 136] (Figure 3). Adenosine receptors are classified in five groups (A_1, A_{2A}, A_{2B}, A_3, and A_4). Subtypes A1 and A3 are dominant in cardiac myocytes, and subtype A2 is dominant in arterial smooth muscle. The A1 subtype activates Gi proteins, and the A2 subtype activates Gs proteins, which subsequently activate adenylyl cyclase and PKA to produce vasodilatation. Based on the similarity between the A1 and A3 receptor subtypes, it was suggested that they have similar functions, although the exact signal transduction pathway from A3 is currently unknown [124, 136]. Son et al. (2005) suggested that the activation of Kir currents by adenosine is mediated only by A3 receptor subtype activation, and consequently by the PKA activation [136]. Park et al. (2005) showed an increase in Kir currents during hypoxia due to activation of PKA in small-diameter coronary arteries [137]. However, it is still unknown if the effects of adenosine on Kir channels are always mediated by the A3 receptor and, as mentioned before, the precise signaling pathway for this receptor is still unknown.

The involvement or modulation of Kir channels by PKG was never demonstrated. However, the NO donor sodium nitroprusside (SNP) activates Kir channels in rat small tail arteries. NO scavenger and Ba^{2+} has been shown to reduce the effect of SNP, suggesting that relaxation by SNP was mediated by Kir channels activation [130, 138].

In conclusion, despite the potential relevance of Kir channel, few studies have addressed crucial aspects of these channels such as molecular structure, regulation by agonists and the role in disease. However some authors think that deeper knowledge about the Kir channels will be important to understand the physiological function of these channels, and their involvement in several cardiovascular pathologies, like hypertension, ischemia and hypertrophy [11, 124, 130].

5. ATP-Sensitive K^+ Channels (K_{ATP})

In 1983 the ATP-sensitive K^+ (K_{ATP}) channels were identified in cardiac muscle. Afterward, these channels have been found in various cells including vascular smooth muscle cells [1, 11]. The K_{ATP} channels confer a degree of metabolic sensitivity to the membrane properties of the cell [139].The K_{ATP} channels are inhibited by cytoplasmic ATP and generally show little voltage sensitivity. In vascular SMC the closing of the K_{ATP} channels leads to vasoconstriction and membrane depolarization. On the contrary, the opening of these channels causes membrane hyperpolarization, decrease in intracellular calcium concentration and vasorelaxation [1, 8, 140]. These channels can be stimulated by vasodilators, while many vasoconstrictors close them, and such modulation represents a major component of their physiological regulation. In arterial smooth muscle cells, several studies have shown that K_{ATP} channels provide a background potassium conductance which is important in the regulation of membrane potential. Then, these channels can modulate arterial tone and blood flow in a several vascular beds [1, 8, 140, 141]. Their activation is closely associated with several pathophysiological responses, including systemic arterial dilation during hypoxia [141, 142], reactive hyperemia in coronary and cerebral circulation [141, 143, 144], and acidosis- and endotoxic shock-induced vasodilation [145]. Moreover, the of K_{ATP} channels seems to play an important role in mediating coronary and cerebral autoregulation (capacity of vascular bed to maintain flow in response to changes in perfusion pressure), and their inhibition disrupts the coronary and cerebral autoregulation [11, 146, 147].

Structure of K_{ATP} Channels

The K_{ATP} channels consist of two different structural subunits: an inwardly rectifying potassium channel (Kir) subunit, which forms the pore (Kir6.x), and a sulfonylurea receptor (SURX) as regulatory subunit [148]. The channel is a tetramer of Kir6.x subunits, which form the K^+ selective pore-forming subunit, surrounded by four SUR proteins [149]. The SUR has high affinity binding sites for sulphonylureas, K^+ channel openers and nucleotides (e.g. ADP). On the other hand, the pore-forming subunits control the magnitude of K^+ flow through the channel and are the sites where ATP binds to inhibit the channel. Based on sequence similarity, Kir6.x is divided into seven subfamilies, including Kir6.1 and Kir6.2. The regulatory sulfonylurea subunit (SUR) is encoded by two different genes, SUR1 and SUR2. SUR2A and SUR2B are splice variants of SUR2 [148]. Different combinations of Kir6.1 or Kir6.2 and SUR1 or SUR2 variants (SUR2A or SUR2B) in cell lines constituted K_{ATP} channels with distinct electrophysiological and pharmacological properties that correspond to the various K_{ATP} channels in native tissues [139]. In vascular smooth muscle the combination Kir6.2/SUR2B is likely the most widespread, although Kir6.1/SUR2B may also be present in this tissue [141, 148].

There is considerable evidence that K_{ATP} channels of vascular SMC can vary in terms of their single channel conductance. The channels clearly fall into two distinct categories: small/medium conductances and large conductances [11, 131]. The values for small/medium-conductance K_{ATP} channels have been reported between 7 and 50 pS and for larger unitary conductances K_{ATP} channels between 130 and 260 pS [11]. The reason for the broad range of single-channel conductances seems to be related with the combination between of the poreforming subunits, Kir6.1 and Kir6.2, to form heterotetramers [150-152]. Channels formed by Kir6.1 or Kir6.2 subunits alone show conductances around 30–35 pS and 70–80 pS respectively, while Kir6.1–Kir6.2 dimers form channels with an intermediate conductance. In vascular SMC the relationship between subunit combination and diversity in the single-channel conductances was demonstrated in some studies. Thus, studies with Kir6.2 knockout mice showed that both, K_{ATP} currents activated by pinacidil in aortic smooth muscle cells and pinacidil relaxations of aortic rings, were unaffected, suggesting that Kir6.1 alone can form the K_{ATP} channel pore in this cells [153]. Despite these studies, further investigations in the molecular basis for K_{ATP} channel diversity in vascular SMC will be essential to understand such diversity in conductance.

Pharmacology of K_{ATP} Channels

The K_{ATP} channels are bloked by antidiabetic sulphonylurea drugs such as glibenclamide and tolbutamide. Glibenclamide is the most frequently used inhibitor in studies performed in arterial smooth muscle cells and tissues. Some artery contractility studies allowed to elucidate that half-inhibitory concentration of glibenclamide oscillates between 20 and 200 nM [1, 11, 154] and for tolbutamide, the half-inhibitory concentration is 350 μM [11]. External Ba^{2+} can also block the K_{ATP} channels, with a half-inhibitory concentration of 100 μM at–80 mV [1, 11].

A number of antihypertensive drugs seem to act through K^+ channels activation and have been designated as K^+ channels openers. These openers exhibit an extreme chemical diversity and comprise a number of different structural classes such as the benzopyrans, cyanoguanidines, thioformamides, benzothiadiazines, and pyridyl nitrates. More recently, a second generation of these drugs was synthesized that include cyclobutenediones, dihydropyridine related structures, and tertiary carbinols (see reviews [155, 156].

Regulation of K_{ATP} Channels

Vascular smooth muscle K_{ATP} channels are regulated by a wide range of vasodilators and vasoconstrictors (Figure 4). Several studies have shown that some vasoconstrictors can inhibit a variety of potassium channels, including K_{ATP} channels in the vascular SMC.

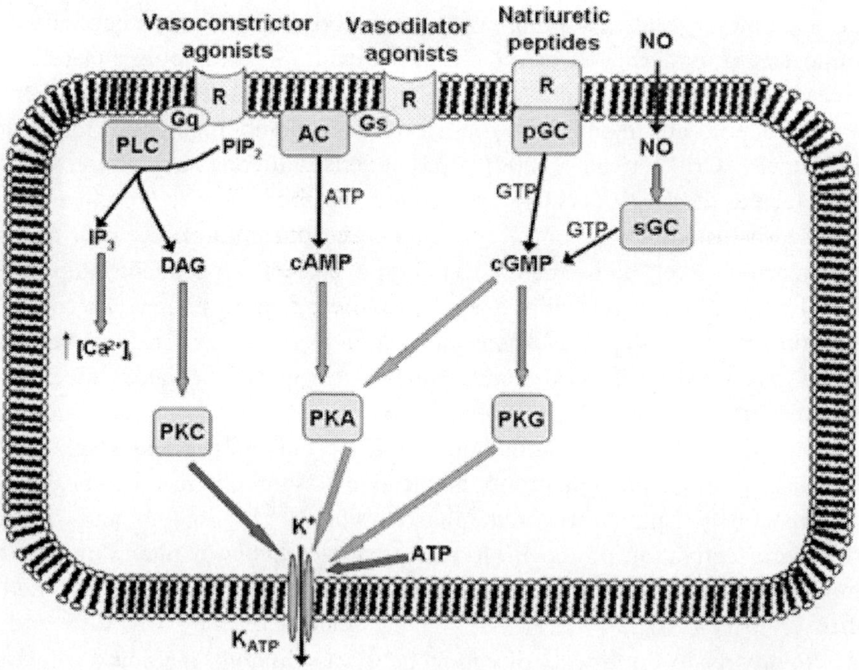

Figure 4. Mechanisms for K_{ATP} channels modulation in vascular smooth muscle cells. The K_{ATP} channels is regulated by vasoconstrictors, vasodilators and cell metabolism. Green arrows means stimulation and red arrows means inhibition. AC, adenylate cyclase; pGC, particulate guanylate cyclase; sGC, soluble guanylate cyclase; NO, nitric oxide; DAG, diacylglycerol ; IP_3, inositol 1,4,5-triphosphate; R, receptor; G, G-protein; cAMP, cyclic adenosine 3',5'-monophosphate; cGMP, cyclic guanosine 3',5'-monophosphate; ATP, adenosine 5'-triphosphate; GTP, guanosine 5'-triphosphate; PKA, cAMP-dependent protein kinase; PKG, cGMP-dependent protein kinase; PKC, protein kinase C.

These vasoconstrictors include angiotensin II [157], endothelin [158], vasopressin [159], serotonin [160], phenylephrine [160], neuropeptide Y [160] and histamine [160]. Several of these vasoconstrictors inhibit K_{ATP} channels through PKC activation (primarily Ca^{2+}-independent PKCε) [8, 11, 141].

Furthermore, vasoconstrictors may inhibit K_{ATP} channels by inhibiting PKA activity [161]. PKC has been shown to modulate the activity of cardiac K_{ATP} channels by phosphorylation of threonine-180 in Kir6.2, and this consensus site is conserved at T190 on vascular SMC in Kir6.1. Previous studies also showed that calcineurin, a Ca^{2+}-dependent protein phosphatase, can modulate K_{ATP} channel activity in isolated VSM cells [162]. Channel activity is inhibited when the intracellular Ca^{2+} concentration increases within the nanomolar range. Furthermore, inhibition of calcineurin activity was proposed to underlie cases of life-threatening syndrome associated with excessive K_{ATP} channel activation in response to drugs. However, neither the site of action nor the calcineurin isoform involved in these effects were elucidated. Orie et al. (2009) used a heterologous expression system (transfected HEK-293 cells) to explore the molecular mechanism of calcineurin modulation on K_{ATP} channels. These authors showed that calcineurin is likely to be the major player in the Ca^{2+}-dependent inhibition of Kir6.1/SUR2B and that this phosphatase suppresses channel activity by antagonizing PKA-dependent phosphorylation [163]. However, the elucidation of the precise physiological significance of this novel regulatory mechanism requires further studies [164]. Recent studies also showed another important pathway involved in the modulation of K_{ATP} channels. Haba et al. (2010) demonstrated that phenylephrine, a selective α1-adrenergic receptor agonist, simultaneously augmented Akt phosphorylation at Ser473 and Thr308. Therefore, activation of the PI3K-Akt pathway seems to play a role in the impairment of K_{ATP} channel function in vascular smooth muscle exposed to phenylephrine [165].

Relatively to the effects of endogenous vasodilators, including calcitonin gene-related peptide (CGRP), β-adrenoceptor agonists and adenosine, several woks have shown that these vasodilators increase K_{ATP} channel activity. Several of them are coupled to Gs receptors, which stimulate adenylyl cyclase and increase intracellular levels of cAMP. These increase can activate either PKA or PKG. Recent data with cloned K_{ATP} channels suggest that channel activity is increased by PKA-dependent phosphorylation. This phosphorylation occurs in the pore-forming and in the regulatory subunits [166]. Even in the absence of vasodilators arterial K_{ATP} channels are subject to a tonic PKA-dependent activation, which arises from sustained cAMP production originating from basal adenylyl cyclase turnover. Recently has been show that vasodilators activate the vascular K_{ATP} channel through direct phosphorylation of the channel protein by PKA in the Ser-1387 [167]. However, it is unclear how phosphorylation at this residue of SUR2B leads to the channel activation.

Concerning cGMP, only a few studies have shown that PKG can activate K_{ATP} channels in certain types of vascular smooth muscle [11]. Some studies showed that nitric oxide, an activator of soluble guanylyl cyclase, relaxes vascular smooth muscle by several mechanisms, one of which is due to activation of K_{ATP} channels [168, 169]. Kubo et al. (1994) showed that atrial natriuretic peptide (ANP), an activator of particulate guanyly cyclase, increases the intracellular cGMP and leads to the activation of K_{ATP} channels in cultured vascular SMC [170]. The pathway involves PKG activation, but some authors also proposed a cross-activation of PKA by cGMP as a possible mechanism for K_{ATP} channels activation [131]. However, recent studies showed that the PKG-mediated phosphorylation exerts dual regulation in the function of K_{ATP} channels. The Kir6.2/SUR1 channels were stimulated after increasing cGMP (by PDE5 inhibition with zaprinast) and this stimulation was abolished by inhibition of PKG, suggesting a stimulatory role of cGMP/PKG signaling in regulating the function of neuronal K_{ATP} channels. In contrast, direct application of purified PKG suppressed rather than activated Kir6.2/SUR1 channels, while tetrameric channels expressed without the SUR subunit were not modulated by zaprinast or purified PKG. Thus, PKG exerts dual functional regulation of neuronal K_{ATP} channels in a SUR subunit-dependent manner.I [171]. In this sense, is important to further analyze the involvement of the PKG in the modulation of the K_{ATP} channels.

Conclusions and Future Perspectives

The K^+ channels play a key role in the determination and/or modulation of the membrane potential. Several vasodilators and vasoconstrictors can act as modulators of these channels. In general, K^+ channels may regulate the vessels contractility tone by altering the cellular membrane potential. It will be important to further investigate the pathways by which these channels can be regulated, and the interaction between the different K^+ channels. These studies involve the determination of the channel structure, the role of the different subunits and the molecular basis of the regulation of the channel activity by the different intracellular pathways. Also, the localization of the different channel types will be important to analyze their role at different vascular levels. An increased knowledge in these matters will also allow the development of new specific drugs to inhibit and activate theses channels that could be useful as therapeutic drugs or pharmacological tolls.

References

[1] Nelson MT, Quayle JM. (1995) Physiological roles and properties of potassium channels in arterial smooth muscle. *Am. J. Physiol.*, 268, C799-822.

[2] Brayden JE, Nelson MT. (1992) Regulation of arterial tone by activation of calcium-dependent potassium channels. *Science*, 256, 532-5.

[3] Harder DR. (1984) Pressure-dependent membrane depolarization in cat middle cerebral artery. *Circ. Res.*, 55, 197-202.

[4] Harder DR, Gilbert R, Lombard JH. (1987) Vascular muscle cell depolarization and activation in renal arteries on elevation of transmural pressure. *Am. J. Physiol.*, 253, F778-81.

[5] Jackson WF. (2000) Ion channels and vascular tone. Hypertension, 35, 173-8.

[6] Kitamura K, Yamazaki J. (2001) Chloride channels and their functional roles in smooth muscle tone in the vasculature. *Jpn J. Pharmacol.*, 85, 351-7.

[7] Chipperfield AR, Harper AA. (2000) Chloride in smooth muscle. *Prog. Biophys Mol. Biol.*, 74, 175-221.

[8] Standen NB, Quayle JM. (1998) K+ channel modulation in arterial smooth muscle. *Acta Physiol.Scand*, 164, 549-57.

[9] Burg ED, Remillard CV, Yuan JX. (2008) Potassium channels in the regulation of pulmonary artery smooth muscle cell proliferation and apoptosis: pharmacotherapeutic implications. *Br. J. Pharmacol.,* 153 Suppl 1, S99-S111.

[10] Sobey CG. (2001) Potassium channel function in vascular disease. *Arterioscler. Thromb. Vasc. Biol.*, 21, 28-38.

[11] Ko EA, Han J, Jung ID, Park WS. (2008) Physiological roles of K+ channels in vascular smooth muscle cells. *J. Smooth Muscle Res.*, 44, 65-81.

[12] Ko EA, Park WS, Firth AL, Kim N, Yuan JX, Han J. (2010) Pathophysiology of voltage-gated K+ channels in vascular smooth muscle cells: modulation by protein kinases. *Prog. Biophys. Mol. Biol.*, 103, 95-101.

[13] Park WS, Firth AL, Han J, Ko EA. (2010) Patho-, physiological roles of voltage-dependent K+ channels in pulmonary arterial smooth muscle cells. *J. Smooth Muscle Res.*, 46, 89-105.

[14] Amberg GC, Koh SD, Imaizumi Y, Ohya S, Sanders KM. (2003) A-type potassium currents in smooth muscle. *Am. J. Physiol. Cell Physiol.*, 284, C583-95.

[15] Gordienko DV, Clausen C, Goligorsky MS. (1994) Ionic currents and endothelin signaling in smooth muscle cells from rat renal resistance arteries. *Am. J. Physiol.,* 266, F325-41.

[16] Clapp LH, Gurney AM. (1991) Outward currents in rabbit pulmonary artery cells dissociated with a new technique. *Exp. Physiol.,* 76, 677-93.

[17] Cairrao E, Morais C, Santos-Silva AJ, Verde I. (2010) PKG is involved in testosterone-induced vasorelaxation of human umbilical artery. *Eur. J. Pharmacol.,* 640, 94-101.

[18] Bonnet S, Archer SL. (2007) Potassium channel diversity in the pulmonary arteries and pulmonary veins: implications for regulation of the pulmonary vasculature in health and during pulmonary hypertension. *Pharmacol. Ther*, 115, 56-69.

[19] Tammaro P, Smith AL, Hutchings SR, Smirnov SV. (2004) Pharmacological evidence for a key role of voltage-gated K+ channels in the function of rat aortic smooth muscle cells. *Br. J. Pharmacol.,* 143, 303-17.

[20] McDaniel SS, Platoshyn O, Yu Y, Sweeney M, Miriel VA, Golovina VA, Krick S, Lapp BR, Wang JY, Yuan JX. (2001) Anorexic effect of K+ channel blockade in mesenteric arterial smooth muscle and intestinal epithelial cells. *J. Appl. Physiol.,* 91, 2322-33.

[21] Papazian DM, Schwarz TL, Tempel BL, Jan YN, Jan LY. (1987) Cloning of genomic and complementary DNA from Shaker, a putative potassium channel gene from Drosophila. *Science*, 237, 749-53.

[22] Wulff H, Castle NA, Pardo LA. (2009) Voltage-gated potassium channels as therapeutic targets. *Nat. Rev. Drug Discov.*, 8, 982-1001.

[23] Long SB, Campbell EB, Mackinnon R. (2005) Crystal structure of a mammalian voltage-dependent Shaker family K+ channel. *Science*, 309, 897-903.

[24] Long SB, Campbell EB, Mackinnon R. (2005) Voltage sensor of Kv1.2: structural basis of electromechanical coupling. *Science*, 309, 903-8.

[25] Doyle DA, Morais Cabral J, Pfuetzner RA, Kuo A, Gulbis JM, Cohen SL, Chait BT, MacKinnon R. (1998) The structure of the potassium

channel: molecular basis of K+ conduction and selectivity. *Science*, 280, 69-77.

[26] Norton RS, Gulbis JM. (2010) Potassium channel gating: not an open and shut case. *Proc. Natl. Acad. Sci. USA*, 107, 7623-4.

[27] Snyders DJ. (1999) Structure and function of cardiac potassium channels. *Cardiovasc. Res.,* 42, 377-90.

[28] MacKinnon R. (2003) Potassium channels. *FEBS Lett*, 555, 62-5.

[29] Elliott DJ, Neale EJ, Aziz Q, Dunham JP, Munsey TS, Hunter M, Sivaprasadarao A. (2004) Molecular mechanism of voltage sensor movements in a potassium channel. *EMBO J.,* 23, 4717-26.

[30] Haitin Y, Yisharel I, Malka E, Shamgar L, Schottelndreier H, Peretz A, Paas Y, Attali B. (2008) S1 constrains S4 in the voltage sensor domain of Kv7.1 K+ channels. *PLoS One*, 3, e1935.

[31] Starace DM, Bezanilla F. (2004) A proton pore in a potassium channel voltage sensor reveals a focused electric field. *Nature*, 427, 548-53.

[32] Ahern CA, Horn R. (2004) Stirring up controversy with a voltage sensor paddle. *Trends Neurosci.*, 27, 303-7.

[33] Ruta V, Chen J, MacKinnon R. (2005) Calibrated measurement of gating-charge arginine displacement in the KvAP voltage-dependent K+ channel. *Cell,* 123, 463-75.

[34] Coetzee WA, Amarillo Y, Chiu J, Chow A, Lau D, McCormack T, Moreno H, Nadal MS, Ozaita A, Pountney D, Saganich M, Vega-Saenz de Miera E, Rudy B. (1999) Molecular diversity of K+ channels. *Ann.* N. *Y Acad. Sci.,* 868, 233-85.

[35] Cox RH. (2005) Molecular determinants of voltage-gated potassium currents in vascular smooth muscle. *Cell Biochem. Biophys*., 42, 167-95.

[36] Pongs O, Schwarz JR. (2010) Ancillary subunits associated with voltage-dependent K+ channels. *Physiol. Rev*., 90, 755-96.

[37] Xu C, Tang G, Lu Y, Wang R. (2000) Molecular basis of voltage-dependent delayed rectifier K+ channels in smooth muscle cells from rat tail artery. *Life Sci*, 66, 2023-33.

[38] Xu C, Lu Y, Tang G, Wang R. (1999) Expression of voltage-dependent K(+) channel genes in mesenteric artery smooth muscle cells. *Am. J. Physiol.,* 277, G1055-63.

[39] Overturf KE, Russell SN, Carl A, Vogalis F, Hart PJ, Hume JR, Sanders KM, Horowitz B. (1994) Cloning and characterization of a Kv1.5 delayed rectifier K+ channel from vascular and visceral smooth muscles. *Am. J. Physiol*., 267, C1231-8.

[40] Immke D, Wood M, Kiss L, Korn SJ. (1999) Potassium-dependent changes in the conformation of the Kv2.1 potassium channel pore. *J. Gen. Physiol.*, 113, 819-36.

[41] Grissmer S, Nguyen AN, Aiyar J, Hanson DC, Mather RJ, Gutman GA, Karmilowicz MJ, Auperin DD, Chandy KG. (1994) Pharmacological characterization of five cloned voltage-gated K+ channels, types Kv1.1, 1.2, 1.3, 1.5, and 3.1, stably expressed in mammalian cell lines. *Mol. Pharmacol.*, 45, 1227-34.

[42] Garcia-Calvo M, Leonard RJ, Novick J, Stevens SP, Schmalhofer W, Kaczorowski GJ, Garcia ML. (1993) Purification, characterization, and biosynthesis of margatoxin, a component of Centruroides margaritatus venom that selectively inhibits voltage-dependent potassium channels. *J. Biol. Chem.*, 268, 18866-74.

[43] Nelson MT, Cheng H, Rubart M, Santana LF, Bonev AD, Knot HJ, Lederer WJ. (1995) Relaxation of arterial smooth muscle by calcium sparks. *Science*, 270, 633-7.

[44] Jonas EA, Kaczmarek LK. (1996) Regulation of potassium channels by protein kinases. *Curr. Opin. Neurobiol.*, 6, 318-23.

[45] Herzig S, Neumann J. (2000) Effects of serine/threonine protein phosphatases on ion channels in excitable membranes. *Physiol. Rev.,* 80, 173-210.

[46] Costa RS, Assreuy J. (2005) Multiple potassium channels mediate nitric oxide-induced inhibition of rat vascular smooth muscle cell proliferation. *Nitric. Oxide*, 13, 145-51.

[47] Wettschureck N, Moers A, Offermanns S. (2004) Mouse models to study G-protein-mediated signaling. *Pharmacol. Ther.*, 101, 75-89.

[48] Sunahara RK, Taussig R. (2002) Isoforms of mammalian adenylyl cyclase: multiplicities of signaling. *Mol. Interv.*, 2, 168-84.

[49] Pierce KL, Lefkowitz RJ. (2001) Classical and new roles of beta-arrestins in the regulation of G-protein-coupled receptors. *Nat. Rev. Neurosci.,* 2, 727-33.

[50] Suh PG, Park JI, Manzoli L, Cocco L, Peak JC, Katan M, Fukami K, Kataoka T, Yun S, Ryu SH. (2008) Multiple roles of phosphoinositide-specific phospholipase C isozymes. *BMB Rep.*, 41, 415-34.

[51] Dixon BS, Sharma RV, Dickerson T, Fortune J. (1994) Bradykinin and angiotensin II: activation of protein kinase C in arterial smooth muscle. *Am. J. Physiol.*, 266, C1406-20.

[52] Lee HC, Cai JJ, Yu H. (1994) Effect of protein kinase C on cyclic 3',5'-adenosine monophosphate-dependent phosphodiesterase in hypertorphic

cardiomyopathic hamster hearts. *J. Pharmacol. Exp. Ther.*, 270, 1171-1176.

[53] Steinberg SF. (2008) Structural basis of protein kinase C isoform function. *Physiol. Rev.,* 88, 1341-78.

[54] Cogolludo A, Moreno L, Lodi F, Frazziano G, Cobeno L, Tamargo J, Perez-Vizcaino F. (2006) Serotonin inhibits voltage-gated K+ currents in pulmonary artery smooth muscle cells: role of 5-HT2A receptors, caveolin-1, and KV1.5 channel internalization. *Circ. Res.,* 98, 931-8.

[55] Cogolludo A, Moreno L, Bosca L, Tamargo J, Perez-Vizcaino F. (2003) Thromboxane A2-induced inhibition of voltage-gated K+ channels and pulmonary vasoconstriction: role of protein kinase Czeta. *Circ. Res.*, 93, 656-63.

[56] Shimoda LA, Sylvester JT, Sham JS. (1998) Inhibition of voltage-gated K+ current in rat intrapulmonary arterial myocytes by endothelin-1. *Am. J. Physiol.*, 274, L842-53.

[57] Shimoda LA, Sylvester JT, Booth GM, Shimoda TH, Meeker S, Undem BJ, Sham JS. (2001) Inhibition of voltage-gated K(+) currents by endothelin-1 in human pulmonary arterial myocytes. *Am. J. Physiol* Lung *Cell Mol. Physiol.*, 281, L1115-22.

[58] Clement-Chomienne O, Walsh MP, Cole WC. (1996) Angiotensin II activation of protein kinase C decreases delayed rectifier K+ current in rabbit vascular myocytes. *J. Physiol.*, 495 (Pt 3), 689-700.

[59] Hayabuchi Y, Standen NB, Davies NW. (2001) Angiotensin II inhibits and alters kinetics of voltage-gated K(+) channels of rat arterial smooth muscle. *Am. J. Physiol. Heart Circ. Physiol.*, 281, H2480-9.

[60] Kwak YG, Navarro-Polanco RA, Grobaski T, Gallagher DJ, Tamkun MM. (1999) Phosphorylation is required for alteration of kv1.5 K(+) channel function by the Kvbeta1.3 subunit. *J. Biol. Chem.*, 274, 25355-61.

[61] Gong J, Xu J, Bezanilla M, van Huizen R, Derin R, Li M. (1999) Differential stimulation of PKC phosphorylation of potassium channels by ZIP1 and ZIP2. *Science*, 285, 1565-9.

[62] Koh SD, Sanders KM, Carl A. (1996) Regulation of smooth muscle delayed rectifier K+ channels by protein kinase A. *Pflugers Arch.*, 432, 401-12.

[63] Aiello EA, Walsh MP, Cole WC. (1995) Phosphorylation by protein kinase A enhances delayed rectifier K+ current in rabbit vascular smooth muscle cells. *Am. J. Physiol.,* 268, H926-34.

[64] Aiello EA, Malcolm AT, Walsh MP, Cole WC. (1998) Beta-adrenoceptor activation and PKA regulate delayed rectifier K+ channels of vascular smooth muscle cells. *Am. J. Physiol.*, 275, H448-59.

[65] Huang XY, Morielli AD, Peralta EG. (1994) Molecular basis of cardiac potassium channel stimulation by protein kinase A. *Proc. Natl. Acad. Sci. USA*, 91, 624-8.

[66] Schrader LA, Anderson AE, Mayne A, Pfaffinger PJ, Sweatt JD. (2002) PKA modulation of Kv4.2-encoded A-type potassium channels requires formation of a supramolecular complex. *J. Neurosci.*, 22, 10123-33.

[67] Levin G, Chikvashvili D, Singer-Lahat D, Peretz T, Thornhill WB, Lotan I. (1996) Phosphorylation of a K+ channel alpha subunit modulates the inactivation conferred by a beta subunit. Involvement of cytoskeleton. *J. Biol. Chem.*, 271, 29321-8.

[68] Yuan XJ, Tod ML, Rubin LJ, Blaustein MP. (1996) NO hyperpolarizes pulmonary artery smooth muscle cells and decreases the intracellular Ca2+ concentration by activating voltage-gated K+ channels. *Proc. Natl. Acad. Sci. USA*, 93, 10489-94.

[69] Zhao YJ, Wang J, Rubin LJ, Yuan XJ. (1997) Inhibition of K(V) and K(Ca) channels antagonizes NO-induced relaxation in pulmonary artery. *Am. J. Physiol.*, 272, H904-12.

[70] Sobey CG, Faraci FM. (1999) Inhibitory effect of 4-aminopyridine on responses of the basilar artery to nitric oxide. *Br. J. Pharmacol.*, 126, 1437-43.

[71] Tanaka Y, Tang G, Takizawa K, Otsuka K, Eghbali M, Song M, Nishimaru K, Shigenobu K, Koike K, Stefani E, Toro L. (2006) Kv channels contribute to nitric oxide- and atrial natriuretic peptide-induced relaxation of a rat conduit artery. *J. Pharmacol. Exp. Ther.*, 317, 341-54.

[72] Dong H, Waldron GJ, Cole WC, Triggle CR. (1998) Roles of calcium-activated and voltage-gated delayed rectifier potassium channels in endothelium-dependent vasorelaxation of the rabbit middle cerebral artery. *Br. J. Pharmacol.*, 123, 821-32.

[73] Heaps CL, Jeffery EC, Laine GA, Price EM, Bowles DK. (2008) Effects of exercise training and hypercholesterolemia on adenosine activation of voltage-dependent K+ channels in coronary arterioles. *J. Appl. Physiol.*, 105, 1761-71.

[74] Luykenaar KD, Brett SE, Wu BN, Wiehler WB, Welsh DG. (2004) Pyrimidine nucleotides suppress KDR currents and depolarize rat cerebral arteries by activating Rho kinase. *Am. J. Physiol Heart Circ.* Physiol, 286, H1088-100.

[75] Cachero TG, Morielli AD, Peralta EG. (1998) The small GTP-binding protein RhoA regulates a delayed rectifier potassium channel. *Cell*, 93, 1077-85.

[76] Sergeant GP, Ohya S, Reihill JA, Perrino BA, Amberg GC, Imaizumi Y, Horowitz B, Sanders KM, Koh SD. (2005) Regulation of Kv4.3 currents by Ca2+/calmodulin-dependent protein kinase II. *Am. J. Physiol Cell Physiol.,* 288, C304-13.

[77] Koh SD, Perrino BA, Hatton WJ, Kenyon JL, Sanders KM. (1999) Novel regulation of the A-type K+ current in murine proximal colon by calcium-calmodulin-dependent protein kinase II. *J. Physiol.*, 517 (Pt 1), 75-84.

[78] Tessier S, Karczewski P, Krause EG, Pansard Y, Acar C, Lang-Lazdunski M, Mercadier JJ, Hatem SN. (1999) Regulation of the transient outward K(+) current by Ca(2+)/calmodulin-dependent protein kinases II in human atrial myocytes. *Circ Res*, 85, 810-9.

[79] Roeper J, Lorra C, Pongs O. (1997) Frequency-dependent inactivation of mammalian A-type K+ channel KV1.4 regulated by Ca2+/calmodulin-dependent protein kinase. *J. Neurosci.*, 17, 3379-91.

[80] Storey NM, O'Bryan JP, Armstrong DL. (2002) Rac and Rho mediate opposing hormonal regulation of the ether-a-go-go-related potassium channel. *Curr. Biol.*, 12, 27-33.

[81] Marty A. (1981) Ca-dependent K channels with large unitary conductance in chromaffin cell membranes. *Nature,* 291, 497-500.

[82] Gardos G. (1958) The function of calcium in the potassium permeability of human erythrocytes. *Biochim. Biophys. Acta*, 30, 653-4.

[83] Blatz AL, Magleby KL. (1986) Single apamin-blocked Ca-activated K+ channels of small conductance in cultured rat skeletal muscle. *Nature*, 323, 718-20.

[84] Ghatta S, Nimmagadda D, Xu X, O'Rourke ST. (2006) Large-conductance, calcium-activated potassium channels: structural and functional implications. *Pharmacol. Ther.*, 110, 103-16.

[85] Brenner R, Perez GJ, Bonev AD, Eckman DM, Kosek JC, Wiler SW, Patterson AJ, Nelson MT, Aldrich RW. (2000) Vasoregulation by the beta1 subunit of the calcium-activated potassium channel. *Nature*, 407, 870-6.

[86] Eichhorn B, Dobrev D. (2007) Vascular large conductance calcium-activated potassium channels: functional role and therapeutic potential. *Naunyn Schmiedebergs Arch. Pharmacol.*, 376, 145-55.

[87] Lee US, Cui J. (2010) BK channel activation: structural and functional insights. *Trends Neurosci.*, 33, 415-23.

[88] Kohler R. (2010) Single-nucleotide polymorphisms in vascular Ca2+-activated K+-channel genes and cardiovascular disease. *Pflugers Arch*, 460, 343-51.

[89] Gollasch M, Lohn M, Furstenau M, Nelson MT, Luft FC, Haller H. (2000) Ca2+ channels, 'quantized' Ca2+ release, and differentiation of myocytes in the cardiovascular system. *J. Hypertens*, 18, 989-98.

[90] Perez GJ, Bonev AD, Patlak JB, Nelson MT. (1999) Functional coupling of ryanodine receptors to KCa channels in smooth muscle cells from rat cerebral arteries. *J. Gen Physiol.*, 113, 229-38.

[91] Hill MA, Yang Y, Ella SR, Davis MJ, Braun AP. (2010) Large conductance, Ca2+-activated K+ channels (BKCa) and arteriolar myogenic signaling. *FEBS Lett*, 584, 2033-42.

[92] Ledoux J, Werner ME, Brayden JE, Nelson MT. (2006) Calcium-activated potassium channels and the regulation of vascular tone. *Physiology* (Bethesda), 21, 69-78.

[93] Cui J, Yang H, Lee US. (2009) Molecular mechanisms of BK channel activation. Cell Mol Life Sci, 66, 852-75.

[94] Waldron GJ, Cole WC. (1999) Activation of vascular smooth muscle K+ channels by endothelium-derived relaxing factors. *Clin. Exp. Pharmacol Physiol.*, 26, 180-4.

[95] Jiang Y, Lee A, Chen J, Cadene M, Chait BT, MacKinnon R. (2002) Crystal structure and mechanism of a calcium-gated potassium channel. *Nature*, 417, 515-22.

[96] Yuan P, Leonetti MD, Pico AR, Hsiung Y, MacKinnon R. (2010) Structure of the human BK channel Ca2+-activation apparatus at 3.0 A resolution. *Science*, 329, 182-6.

[97] Wallner M, Meera P, Ottolia M, Kaczorowski GJ, Latorre R, Garcia ML, Stefani E, Toro L. (1995) Characterization of and modulation by a beta-subunit of a human maxi KCa channel cloned from myometrium. *Receptors Channels*, 3, 185-99.

[98] Tanaka Y, Koike K, Toro L. (2004) MaxiK channel roles in blood vessel relaxations induced by endothelium-derived relaxing factors and their molecular mechanisms. *J. Smooth Muscle Res.*, 40, 125-53.

[99] Jiang Z, Wallner M, Meera P, Toro L. (1999) Human and rodent MaxiK channel beta-subunit genes: cloning and characterization. *Genomics*, 55, 57-67.

[100] Liegeois JF, Mercier F, Graulich A, Graulich-Lorge F, Scuvee-Moreau J, Seutin V. (2003) Modulation of small conductance calcium-activated potassium (SK) channels: a new challenge in medicinal chemistry. *Curr. Med. Chem.*, 10, 625-47.

[101] Lu R, Alioua A, Kumar Y, Eghbali M, Stefani E, Toro L. (2006) MaxiK channel partners: physiological impact. *J. Physiol.*, 570, 65-72.

[102] Alioua A, Mahajan A, Nishimaru K, Zarei MM, Stefani E, Toro L. (2002) Coupling of c-Src to large conductance voltage- and Ca2+-activated K+ channels as a new mechanism of agonist-induced vasoconstriction. *Proc. Natl. Acad. Sci. USA,* 99, 14560-5.

[103] Dick GM, Sanders KM. (2001) (Xeno)estrogen sensitivity of smooth muscle BK channels conferred by the regulatory beta1 subunit: a study of beta1 knockout mice. *J. Biol. Chem.*, 276, 44835-40.

[104] White RE. (2002) Estrogen and vascular function. *Vascul. Pharmacol*, 38, 73-80.

[105] Kim JK, Levin ER. (2006) Estrogen signaling in the cardiovascular system. *Nucl. Recept Signal.*, 4, e013.

[106] Zhu N, Eghbali M, Helguera G, Song M, Stefani E, Toro L. (2005) Alternative splicing of Slo channel gene programmed by estrogen, progesterone and pregnancy. *FEBS Lett*, 579, 4856-60.

[107] Cairrao E, Alvarez E, Santos-Silva AJ, Verde I. (2008) Potassium channels are involved in testosterone-induced vasorelaxation of human umbilical artery. *Naunyn Schmiedebergs Arch. Pharmacol.*, 376, 375-83.

[108] White RE, Darkow DJ, Lang JL. (1995) Estrogen relaxes coronary arteries by opening BKCa channels through a cGMP-dependent mechanism. *Circ. Res.,* 77, 936-42.

[109] Geary GG, Krause DN, Duckles SP. (1998) Estrogen reduces myogenic tone through a nitric oxide-dependent mechanism in rat cerebral arteries. *Am. J. Physiol.,* 275, H292-300.

[110] Bolotina VM, Najibi S, Palacino JJ, Pagano PJ, Cohen RA. (1994) Nitric oxide directly activates calcium-dependent potassium channels in vascular smooth muscle. *Nature,* 368, 850-3.

[111] Dick GM, Rossow CF, Smirnov S, Horowitz B, Sanders KM. (2001) Tamoxifen activates smooth muscle BK channels through the regulatory beta 1 subunit. *J. Biol. Chem.*, 276, 34594-9.

[112] Alioua A, Huggins JP, Rousseau E. (1995) PKG-I alpha phosphorylates the alpha-subunit and upregulates reconstituted GKCa channels from tracheal smooth muscle. *Am. J. Physiol.*, 268, L1057-63.

[113] Robertson BE, Schubert R, Hescheler J, Nelson MT. (1993) cGMP-dependent protein kinase activates Ca-activated K channels in cerebral artery smooth muscle cells. *Am. J. Physiol.*, 265, C299-303.

[114] Jaggar JH, Wellman GC, Heppner TJ, Porter VA, Perez GJ, Gollasch M, Kleppisch T, Rubart M, Stevenson AS, Lederer WJ, Knot HJ, Bonev AD, Nelson MT. (1998) Ca2+ channels, ryanodine receptors and Ca2+-activated K+ channels: a functional unit for regulating arterial tone. *Acta Physiol. Scand*, 164, 577-87.

[115] Wellman GC, Nelson MT. (2003) Signaling between SR and plasmalemma in smooth muscle: sparks and the activation of Ca2+-sensitive ion channels. *Cell Calcium*, 34, 211-29.

[116] Wellman GC, Santana LF, Bonev AD, Nelson MT. (2001) Role of phospholamban in the modulation of arterial Ca(2+) sparks and Ca(2+)-activated K(+) channels by cAMP. *Am. J. Physiol. Cell Physiol.*, 281, C1029-37.

[117] Minami K, Fukuzawa K, Nakaya Y. (1993) Protein kinase C inhibits the Ca(2+)-activated K+ channel of cultured porcine coronary artery smooth muscle cells. *Biochem. Biophys. Res. Commun.*, 190, 263-9.

[118] Zhu S, White RE, Barman SA. (2006) Effect of PKC isozyme inhibition on forskolin-induced activation of BKCa channels in rat pulmonary arterial smooth muscle. *Lung,* 184, 89-97.

[119] Barman SA, Zhu S, White RE. (2004) Protein kinase C inhibits BKCa channel activity in pulmonary arterial smooth muscle. *Am. J. Physiol.,* 286, L149-55.

[120] Zhou XB, Arntz C, Kamm S, Motejlek K, Sausbier U, Wang GX, Ruth P, Korth M. (2001) A molecular switch for specific stimulation of the BKCa channel by cGMP and cAMP kinase. *J. Biol. Chem.*, 276, 43239-45.

[121] Nichols CG, Lopatin AN. (1997) Inward rectifier potassium channels. *Annu. Rev. Physiol.*, 59, 171-91.

[122] Lu Z. (2004) Mechanism of rectification in inward-rectifier K+ channels. *Annu. Rev. Physiol.*, 66, 103-29.

[123] Logothetis DE, Jin T, Lupyan D, Rosenhouse-Dantsker A. (2007) Phosphoinositide-mediated gating of inwardly rectifying K(+) channels. *Pflugers Arch*, 455, 83-95.

[124] Park WS, Han J, Earm YE. (2008) Physiological role of inward rectifier K(+) channels in vascular smooth muscle cells. *Pflugers Arch.*, 457, 137-47.

[125] Stanfield PR, Nakajima S, Nakajima Y. (2002) Constitutively active and G-protein coupled inward rectifier K+ channels: Kir2.0 and Kir3.0. *Rev. Physiol. Biochem. Pharmacol.,* 145, 47-179.

[126] Quayle JM, McCarron JG, Brayden JE, Nelson MT. (1993) Inward rectifier K+ currents in smooth muscle cells from rat resistance-sized cerebral arteries. *Am. J. Physiol.*, 265, C1363-70.

[127] Wu BN, Luykenaar KD, Brayden JE, Giles WR, Corteling RL, Wiehler WB, Welsh DG. (2007) Hyposmotic challenge inhibits inward rectifying K+ channels in cerebral arterial smooth muscle cells. *Am. J. Physiol. Heart Circ. Physiol*, 292, H1085-94.

[128] Tennant BP, Cui Y, Tinker A, Clapp LH. (2006) Functional expression of inward rectifier potassium channels in cultured human pulmonary smooth muscle cells: evidence for a major role of Kir2.4 subunits. *J. Membr. Biol.,* 213, 19-29.

[129] Robertson BE, Bonev AD, Nelson MT. (1996) Inward rectifier K+ currents in smooth muscle cells from rat coronary arteries: block by Mg2+, Ca2+, and Ba2+. *Am. J. Physiol.*, 271, H696-705.

[130] Jackson WF. (2005) Potassium channels in the peripheral microcirculation. *Microcirculation*, 12, 113-27.

[131] Quayle JM, Nelson MT, Standen NB. (1997) ATP-sensitive and inwardly rectifying potassium channels in smooth muscle. *Physiol. Rev.,* 77, 1165-232.

[132] Park WS, Kim N, Youm JB, Warda M, Ko JH, Kim SJ, Earm YE, Han J. (2006) Angiotensin II inhibits inward rectifier K+ channels in rabbit coronary arterial smooth muscle cells through protein kinase Calpha. *Biochem. Biophys. Res. Commun.*, 341, 728-35.

[133] Park WS, Han J, Kim N, Youm JB, Joo H, Kim HK, Ko JH, Earm YE. (2005) Endothelin-1 inhibits inward rectifier K+ channels in rabbit coronary arterial smooth muscle cells through protein kinase C. *J. Cardiovasc. Pharmacol.,* 46, 681-9.

[134] Xu X, Rials SJ, Wu Y, Marinchak RA, Kowey PR. (1999) The properties of the inward rectifier potassium currents in rabbit coronary arterial smooth muscle cells. *Pflugers Arch*, 438, 187-94.

[135] Ruppersberg JP. (2000) Intracellular regulation of inward rectifier K+ channels. *Pflugers Arch,* 441, 1-11.

[136] Son YK, Park WS, Ko JH, Han J, Kim N, Earm YE. (2005) Protein kinase A-dependent activation of inward rectifier potassium channels by adenosine in rabbit coronary smooth muscle cells. *Biochem. Biophys. Res Commun.*, 337, 1145-52.

[137] Park WS, Han J, Kim N, Ko JH, Kim SJ, Earm YE. (2005) Activation of inward rectifier K+ channels by hypoxia in rabbit coronary arterial smooth muscle cells. *Am. J. Physiol. Heart Circ. Physiol.,* 289, H2461-7.

[138] Schubert R, Krien U, Wulfsen I, Schiemann D, Lehmann G, Ulfig N, Veh RW, Schwarz JR, Gago H. (2004) Nitric oxide donor sodium nitroprusside dilates rat small arteries by activation of inward rectifier potassium channels. *Hypertension*, 43, 891-6.

[139] Miki T, Seino S. (2005) Roles of KATP channels as metabolic sensors in acute metabolic changes. *J. Mol. Cell Cardiol.*, 38, 917-25.

[140] Quayle JM, Standen NB. (1994) KATP channels in vascular smooth muscle. *Cardiovasc. Res.*, 28, 797-804.

[141] Brayden JE. (2002) Functional roles of KATP channels in vascular smooth muscle. *Clin. Exp. Pharmacol. Physiol.,* 29, 312-6.

[142] Daut J, Maier-Rudolph W, von Beckerath N, Mehrke G, Gunther K, Goedel-Meinen L. (1990) Hypoxic dilation of coronary arteries is mediated by ATP-sensitive potassium channels. *Science*, 247, 1341-4.

[143] Kanatsuka H, Sekiguchi N, Sato K, Akai K, Wang Y, Komaru T, Ashikawa K, Takishima T. (1992) Microvascular sites and mechanisms responsible for reactive hyperemia in the coronary circulation of the beating canine heart. *Circ. Res.*, 71, 912-22.

[144] Bari F, Louis TM, Busija DW. (1998) Effects of ischemia on cerebral arteriolar dilation to arterial hypoxia in piglets. *Stroke,* 29, 222-7; discussion 227-8.

[145] Landry DW, Oliver JA. (1992) The ATP-sensitive K+ channel mediates hypotension in endotoxemia and hypoxic lactic acidosis in dog. *J. Clin. Invest*, 89, 2071-4.

[146] Hong KW, Pyo KM, Lee WS, Yu SS, Rhim BY. (1994) Pharmacological evidence that calcitonin gene-related peptide is implicated in cerebral autoregulation. *Am. J. Physiol.,* 266, H11-6.

[147] Narishige T, Egashira K, Akatsuka Y, Katsuda Y, Numaguchi K, Sakata M, Takeshita A. (1993) Glibenclamide, a putative ATP-sensitive K+ channel blocker, inhibits coronary autoregulation in anesthetized dogs. *Circ. Res.*, 73, 771-6.

[148] Seino S. (2003) Physiology and pathophysiology of K(ATP) channels in the pancreas and cardiovascular system: a review. *J. Diabetes Complications*, 17, 2-5.

[149] Buckley JF, Singer M, Clapp LH. (2006) Role of KATP channels in sepsis. *Cardiovasc. Res.*, 72, 220-30.

[150] Cui Y, Giblin JP, Clapp LH, Tinker A. (2001) A mechanism for ATP-sensitive potassium channel diversity: Functional coassembly of two pore-forming subunits. *Proc. Natl. Acad. Sci. USA*, 98, 729-34.

[151] Pountney DJ, Sun ZQ, Porter LM, Nitabach MN, Nakamura TY, Holmes D, Rosner E, Kaneko M, Manaris T, Holmes TC, Coetzee WA. (2001) Is the molecular composition of K(ATP) channels more complex than originally thought? *J. Mol. Cell Cardiol.*, 33, 1541-6.

[152] Chan KW, Wheeler A, Csanady L. (2008) Sulfonylurea receptors type 1 and 2A randomly assemble to form heteromeric KATP channels of mixed subunit composition. *J. Gen. Physiol.,* 131, 43-58.

[153] Suzuki M, Li RA, Miki T, Uemura H, Sakamoto N, Ohmoto-Sekine Y, Tamagawa M, Ogura T, Seino S, Marban E, Nakaya H. (2001) Functional roles of cardiac and vascular ATP-sensitive potassium channels clarified by Kir6.2-knockout mice. *Circ. Res.,* 88, 570-7.

[154] Wellman GC, Quayle JM, Standen NB. (1998) ATP-sensitive K+ channel activation by calcitonin gene-related peptide and protein kinase A in pig coronary arterial smooth muscle. *J. Physiol.,* 507 (Pt 1), 117-29.

[155] Mannhold R. (2004) KATP channel openers: structure-activity relationships and therapeutic potential. *Med. Res. Rev.*, 24, 213-66.

[156] Mannhold R. (2006) Structure-activity relationships of K(ATP) channel openers. *Curr. Top Med. Chem.*, 6, 1031-47.

[157] Kubo M, Quayle JM, Standen NB. (1997) Angiotensin II inhibition of ATP-sensitive K+ currents in rat arterial smooth muscle cells through protein kinase C. *J. Physiol.*, 503 (Pt 3), 489-96.

[158] Park WS, Ko EA, Han J, Kim N, Earm YE. (2005) Endothelin-1 acts via protein kinase C to block KATP channels in rabbit coronary and pulmonary arterial smooth muscle cells. *J. Cardiovasc. Pharmacol.*, 45, 99-108.

[159] Wakatsuki T, Nakaya Y, Inoue I. (1992) Vasopressin modulates K(+)-channel activities of cultured smooth muscle cells from porcine coronary artery. *Am. J. Physiol., 263*, H491-6.

[160] Bonev AD, Nelson MT. (1996) Vasoconstrictors inhibit ATP-sensitive K+ channels in arterial smooth muscle through protein kinase C. *J. Gen. Physiol.,* 108, 315-23.

[161] Hayabuchi Y, Davies NW, Standen NB. (2001) Angiotensin II inhibits rat arterial KATP channels by inhibiting steady-state protein kinase A activity and activating protein kinase Ce. *J. Physiol.*, 530, 193-205.

[162] Wilson AJ, Jabr RI, Clapp LH. (2000) Calcium modulation of vascular smooth muscle ATP-sensitive K(+) channels: role of protein phosphatase-2B. *Circ. Res.*, 87, 1019-25.

[163] Orie NN, Thomas AM, Perrino BA, Tinker A, Clapp LH. (2009) Ca2+/calcineurin regulation of cloned vascular K ATP channels: crosstalk with the protein kinase A pathway. *Br. J. Pharmacol.*, 157, 554-64.

[164] Tammaro P. (2009) Vascular K ATP channels: dephosphorylation and deactivation. *Br. J. Pharmacol.*, 157, 551-3.

[165] Haba M, Hatakeyama N, Kinoshita H, Teramae H, Azma T, Hatano Y, Matsuda N. (2010) The modulation of vascular ATP-sensitive K+ channel function via the phosphatidylinositol 3-kinase-Akt pathway activated by phenylephrine. *J. Pharmacol. Exp. Ther.*, 334, 673-8.

[166] Quinn KV, Giblin JP, Tinker A. (2004) Multisite phosphorylation mechanism for protein kinase A activation of the smooth muscle ATP-sensitive K+ channel. *Circ. Res*, 94, 1359-66.

[167] Shi Y, Chen X, Wu Z, Shi W, Yang Y, Cui N, Jiang C, Harrison RW. (2008) cAMP-dependent protein kinase phosphorylation produces interdomain movement in SUR2B leading to activation of the vascular KATP channel. *J. Biol. Chem.*, 283, 7523-30.

[168] Murphy ME, Brayden JE. (1995) Nitric oxide hyperpolarizes rabbit mesenteric arteries via ATP-sensitive potassium channels. *J. Physiol*, 486 (Pt 1), 47-58.

[169] Miyoshi H, Nakaya Y, Moritoki H. (1994) Nonendothelial-derived nitric oxide activates the ATP-sensitive K+ channel of vascular smooth muscle cells. *FEBS Lett*, 345, 47-9.

[170] Kubo M, Nakaya Y, Matsuoka S, Saito K, Kuroda Y. (1994) Atrial natriuretic factor and isosorbide dinitrate modulate the gating of ATP-sensitive K+ channels in cultured vascular smooth muscle cells. *Circ. Res.,* 74, 471-6.

[171] Chai Y, Lin YF. (2008) Dual regulation of the ATP-sensitive potassium channel by activation of cGMP-dependent protein kinase. Pflugers Arch, 456, 897-915.

[172] Koschak A, Bugianesi RM, Mitterdorfer J, Kaczorowski GJ, Garcia ML, Knaus HG. (1998) Subunit composition of brain voltage-gated potassium channels determined by hongotoxin-1, a novel peptide derived from Centruroides limbatus venom. *J. Biol. Chem.*, 273, 2639-44.

[173] Rodriguez de la Vega RC, Merino E, Becerril B, Possani LD. (2003) Novel interactions between K+ channels and scorpion toxins. *Trends Pharmacol. Sci.,* 24, 222-7.

[174] Kauferstein S, Huys I, Lamthanh H, Stocklin R, Sotto F, Menez A, Tytgat J, Mebs D. (2003) A novel conotoxin inhibiting vertebrate voltage-sensitive potassium channels. *Toxicon,* 42, 43-52.

[175] Zhang ZH, Rhodes KJ, Childers WE, Argentieri TM, Wang Q. (2004) Disinactivation of N-type inactivation of voltage-gated K channels by an erbstatin analogue. *J. Biol. Chem.*, 279, 29226-30.

[176] Lu Q, Peevey J, Jow F, Monaghan MM, Mendoza G, Zhang H, Wu J, Kim CY, Bicksler J, Greenblatt L, Lin SS, Childers W, Bowlby MR. (2008) Disruption of Kv1.1 N-type inactivation by novel small molecule inhibitors (disinactivators). *Bioorg. Med. Chem.,* 16, 3067-75.

[177] M'Barek S, Mosbah A, Sandoz G, Fajloun Z, Olamendi-Portugal T, Rochat H, Sampieri F, Guijarro JI, Mansuelle P, Delepierre M, De Waard M, Sabatier JM. (2003) Synthesis and characterization of Pi4, a scorpion toxin from Pandinus imperator that acts on K+ channels. *Eur. J. Biochem.*, 270, 3583-92.

[178] Rauer H, Pennington M, Cahalan M, Chandy KG. (1999) Structural conservation of the pores of calcium-activated and voltage-gated potassium channels determined by a sea anemone toxin. *J. Biol. Chem.,* 274, 21885-92.

[179] Schmitz A, Sankaranarayanan A, Azam P, Schmidt-Lassen K, Homerick D, Hansel W, Wulff H. (2005) Design of PAP-1, a selective small molecule Kv1.3 blocker, for the suppression of effector memory T cells in autoimmune diseases. *Mol. Pharmacol.*, 68, 1254-70.

[180] Baell JB, Gable RW, Harvey AJ, Toovey N, Herzog T, Hansel W, Wulff H. (2004) Khellinone derivatives as blockers of the voltage-gated potassium channel Kv1.3: synthesis and immunosuppressive activity. *J. Med. Chem.*, 47, 2326-36.

[181] Harvey AJ, Baell JB, Toovey N, Homerick D, Wulff H. (2006) A new class of blockers of the voltage-gated potassium channel Kv1.3 via modification of the 4- or 7-position of khellinone. *J. Med. Chem.*, 49, 1433-41.

[182] Ren YR, Pan F, Parvez S, Fleig A, Chong CR, Xu J, Dang Y, Zhang J, Jiang H, Penner R, Liu JO. (2008) Clofazimine inhibits human Kv1.3 potassium channel by perturbing calcium oscillation in T lymphocytes. *PLoS One*, 3, e4009.

[183] Yeola SW, Snyders DJ. (1997) Electrophysiological and pharmacological correspondence between Kv4.2 current and rat cardiac transient outward current. *Cardiovasc. Res.*, 33, 540-7.

[184] Gross MF, Beaudoin S, McNaughton-Smith G, Amato GS, Castle NA, Huang C, Zou A, Yu W. (2007) Aryl sulfonamido indane inhibitors of the Kv1.5 ion channel. *Bioorg. Med. Chem. Lett.*, 17, 2849-53.

[185] Wirth KJ, Paehler T, Rosenstein B, Knobloch K, Maier T, Frenzel J, Brendel J, Busch AE, Bleich M. (2003) Atrial effects of the novel K(+)-channel-blocker AVE0118 in anesthetized pigs. *Cardiovasc. Res.*, 60, 298-306.

[186] Wirth KJ, Brendel J, Steinmeyer K, Linz DK, Rutten H, Gogelein H. (2007) In vitro and in vivo effects of the atrial selective antiarrhythmic compound AVE1231. *J. Cardiovasc. Pharmacol.*, 49, 197-206.

[187] Eldstrom J, Wang Z, Xu H, Pourrier M, Ezrin A, Gibson K, Fedida D. (2007) The molecular basis of high-affinity binding of the antiarrhythmic compound vernakalant (RSD1235) to Kv1.5 channels. *Mol. Pharmacol.*, 72, 1522-34.

[188] Regan CP, Kiss L, Stump GL, McIntyre CJ, Beshore DC, Liverton NJ, Dinsmore CJ, Lynch JJ, Jr. (2008) Atrial antifibrillatory effects of structurally distinct IKur blockers 3-[(dimethylamino)methyl]-6-methoxy-2-methyl-4-phenylisoquinolin-1(2H)-one and 2-phenyl-1,1-dipyridin-3-yl-2-pyrrolidin-1-yl-ethanol in dogs with underlying heart failure. *J. Pharmacol. Exp. Ther.*, 324, 322-30.

[189] Wu S, Fluxe A, Sheffer J, Janusz JM, Blass BE, White R, Jackson C, Hedges R, Murawsky M, Fang B, Fadayel GM, Hare M, Djandjighian L. (2006) Discovery and in vitro/in vivo studies of tetrazole derivatives as Kv1.5 blockers. *Bioorg. Med. Chem. Lett,* 16, 6213-8.

[190] Lagrutta A, Wang J, Fermini B, Salata JJ. (2006) Novel, potent inhibitors of human Kv1.5 K+ channels and ultrarapidly activating delayed rectifier potassium current. *J. Pharmacol. Exp. Ther.*, 317, 1054-63.

[191] Ford JW, Milnes JT. (2008) New drugs targeting the cardiac ultra-rapid delayed-rectifier current (I Kur): rationale, pharmacology and evidence for potential therapeutic value. *J. Cardiovasc. Pharmacol.*, 52, 105-20.

[192] Kalman K, Nguyen A, Tseng-Crank J, Dukes ID, Chandy G, Hustad CM, Copeland NG, Jenkins NA, Mohrenweiser H, Brandriff B, Cahalan M, Gutman GA, Chandy KG. (1998) Genomic organization, chromosomal localization, tissue distribution, and biophysical

characterization of a novel mammalian Shaker-related voltage-gated potassium channel, Kv1.7. *J. Biol. Chem.*, 273, 5851-7.

[193] Escoubas P, Diochot S, Celerier ML, Nakajima T, Lazdunski M. (2002) Novel tarantula toxins for subtypes of voltage-dependent potassium channels in the Kv2 and Kv4 subfamilies. *Mol. Pharmacol.*, 62, 48-57.

[194] Schmalz F, Kinsella J, Koh SD, Vogalis F, Schneider A, Flynn ER, Kenyon JL, Horowitz B. (1998) Molecular identification of a component of delayed rectifier current in gastrointestinal smooth muscles. *Am. J. Physiol.,* 274, G901-11.

[195] Diochot S, Schweitz H, Beress L, Lazdunski M. (1998) Sea anemone peptides with a specific blocking activity against the fast inactivating potassium channel Kv3.4. *J. Biol. Chem.,* 273, 6744-9.

[196] Kestler HA, Janko S, Haussler U, Muche R, Hombach V, Hoher M, Wiecha J. (1998) A remark on the high-conductance calcium-activated potassium channel in human endothelial cells. *Res. Exp. Med.* (Berl), 198, 133-43.

[197] Galvez A, Gimenez-Gallego G, Reuben JP, Roy-Contancin L, Feigenbaum P, Kaczorowski GJ, Garcia ML. (1990) Purification and characterization of a unique, potent, peptidyl probe for the high conductance calcium-activated potassium channel from venom of the scorpion Buthus tamulus. *J. Biol. Chem.*, 265, 11083-90.

[198] Giangiacomo KM, Garcia ML, McManus OB. (1992) Mechanism of iberiotoxin block of the large-conductance calcium-activated potassium channel from bovine aortic smooth muscle. *Biochemistry*, 31, 6719-27.

[199] Li G, Cheung DW. (1999) Effects of paxilline on K+ channels in rat mesenteric arterial cells. *Eur. J. Pharmacol.,* 372, 103-7.

[200] Li HF, Chen SA, Wu SN. (2000) Evidence for the stimulatory effect of resveratrol on Ca(2+)-activated K+ current in vascular endothelial cells. *Cardiovasc. Res.,* 45, 1035-45.

[201] Knaus HG, McManus OB, Lee SH, Schmalhofer WA, Garcia-Calvo M, Helms LM, Sanchez M, Giangiacomo K, Reuben JP, Smith AB, 3rd, et al. (1994) Tremorgenic indole alkaloids potently inhibit smooth muscle high-conductance calcium-activated potassium channels. *Biochemistry,* 33, 5819-28.

[202] McGahon MK, Dash DP, Arora A, Wall N, Dawicki J, Simpson DA, Scholfield CN, McGeown JG, Curtis TM. (2007) Diabetes downregulates large-conductance Ca2+-activated potassium beta 1 channel subunit in retinal arteriolar smooth muscle. *Circ. Res.*, 100, 703-11.

[203] Edwards G, Niederste-Hollenberg A, Schneider J, Noack T, Weston AH. (1994) Ion channel modulation by NS 1619, the putative BKCa channel opener, in vascular smooth muscle. *Br. J. Pharmacol.*, 113, 1538-47.

[204] Ahn SC, Seol GH, Kim JA, Suh SH. (2004) Characteristics and a functional implication of Ca2+-activated K+ current in mouse aortic endothelial cells. *Pflugers Arch*, 447, 426-35.

[205] Bychkov R, Burnham MP, Richards GR, Edwards G, Weston AH, Feletou M, Vanhoutte PM. (2002) Characterization of a charybdotoxin-sensitive intermediate conductance Ca2+-activated K+ channel in porcine coronary endothelium: relevance to EDHF. *Br. J. Pharmacol.*, 137, 1346-54.

[206] Sollini M, Frieden M, Beny JL. (2002) Charybdotoxin-sensitive small conductance K(Ca) channel activated by bradykinin and substance P in endothelial cells. *Br. J. Pharmacol.*, 136, 1201-9.

[207] Rusko J, Tanzi F, van Breemen C, Adams DJ. (1992) Calcium-activated potassium channels in native endothelial cells from rabbit aorta: conductance, Ca2+ sensitivity and block. *J. Physiol.*, 455, 601-21.

[208] Catacuzzeno L, Pisconti DA, Harper AA, Petris A, Franciolini F. (2000) Characterization of the large-conductance Ca-activated K channel in myocytes of rat saphenous artery. *Pflugers Arch*, 441, 208-18.

[209] Nilius B, Droogmans G. (2001) Ion channels and their functional role in vascular endothelium. *Physiol. Rev.*, 81, 1415-59.

[210] Calderone V. (2002) Large-conductance, Ca2+-activated K+ channels: function, pharmacology and drugs. *Curr. Med. Chem.*, 9, 1385-95.

[211] Calderone V, Martelli A, Testai L, Martinotti E, Breschi MC. (2007) Functional contribution of the endothelial component to the vasorelaxing effect of resveratrol and NS 1619, activators of the large-conductance calcium-activated potassium channels. *Naunyn Schmiedebergs Arch Pharmacol.*, 375, 73-80.

[212] Xu CQ, Brone B, Wicher D, Bozkurt O, Lu WY, Huys I, Han YH, Tytgat J, Van Kerkhove E, Chi CW. (2004) BmBKTx1, a novel Ca2+-activated K+ channel blocker purified from the Asian scorpion Buthus martensi Karsch. *J. Biol. Chem.*, 279, 34562-9.

[213] Yao J, Chen X, Li H, Zhou Y, Yao L, Wu G, Zhang N, Zhou Z, Xu T, Wu H, Ding J. (2005) BmP09, a "long chain" scorpion peptide blocker of BK channels. *J. Biol. Chem.*, 280, 14819-28.

[214] Garcia-Valdes J, Zamudio FZ, Toro L, Possani LD. (2001) Slotoxin, alphaKTx1.11, a new scorpion peptide blocker of MaxiK channels that

differentiates between alpha and alpha+beta (beta1 or beta4) complexes. *FEBS Lett,* 505, 369-73.

[215] McKay MC, Dworetzky SI, Meanwell NA, Olesen SP, Reinhart PH, Levitan IB, Adelman JP, Gribkoff VK. (1994) Opening of large-conductance calcium-activated potassium channels by the substituted benzimidazolone NS004. *J. Neurophysiol.*, 71, 1873-82.

[216] Xu X, Tsai TD, Wang J, Lee EW, Lee KS. (1994) Modulation of three types of K+ currents in canine coronary artery smooth muscle cells by NS-004, or 1-(2'-hydroxy-5'-chlorophenyl)-5-trifluoromethyl-2(3H) benzimidazolone. *J. Pharmacol. Exp. Ther.*, 271, 362-9.

[217] Hu S, Kim HS, Fink CA. (1995) Differential effects of the BKCa channel openers NS004 and NS1608 in porcine coronary arterial cells. *Eur. J. Pharmacol.*, 294, 357-60.

[218] Olesen SP, Munch E, Moldt P, Drejer J. (1994) Selective activation of Ca(2+)-dependent K+ channels by novel benzimidazolone. *Eur. J. Pharmacol.,* 251, 53-9.

[219] McManus OB, Harris GH, Giangiacomo KM, Feigenbaum P, Reuben JP, Addy ME, Burka JF, Kaczorowski GJ, Garcia ML. (1993) An activator of calcium-dependent potassium channels isolated from a medicinal herb. *Biochemistry,* 32, 6128-33.

[220] Tanaka Y, Meera P, Song M, Knaus HG, Toro L. (1997) Molecular constituents of maxi KCa channels in human coronary smooth muscle: predominant alpha + beta subunit complexes. *J. Physiol.,* 502 (Pt 3), 545-57.

[221] Kim HI, Kim TH, Song JH. (2005) Resveratrol inhibits Na+ currents in rat dorsal root ganglion neurons. *Brain Res,* 1045, 134-41.

[222] Nagaoka T, Hein TW, Yoshida A, Kuo L. (2007) Resveratrol, a component of red wine, elicits dilation of isolated porcine retinal arterioles: role of nitric oxide and potassium channels. *Invest Ophthalmol. Vis. Sci.*, 48, 4232-9.

[223] Grover GJ, D'Alonzo AJ, Sleph PG, Dzwonczyk S, Hess TA, Darbenzio RB. (1994) The cardioprotective and electrophysiological effects of cromakalim are attenuated by meclofenamate through a cyclooxygenase-independent mechanism. *J. Pharmacol. Exp. Ther.*, 269, 536-40.

[224] Malykhina AP, Shoeb F, Akbarali HI. (2002) Fenamate-induced enhancement of heterologously expressed HERG currents in Xenopus oocytes. *Eur. J. Pharmacol.*, 452, 269-77.

[225] White MM, Aylwin M. (1990) Niflumic and flufenamic acids are potent reversible blockers of Ca2(+)-activated Cl- channels in Xenopus oocytes. *Mol. Pharmacol.*, 37, 720-4.

[226] Gogelein H, Dahlem D, Englert HC, Lang HJ. (1990) Flufenamic acid, mefenamic acid and niflumic acid inhibit single nonselective cation channels in the rat exocrine pancreas. *FEBS Lett*, 268, 79-82.

[227] Lee YT, Wang Q. (1999) Inhibition of hKv2.1, a major human neuronal voltage-gated K+ channel, by meclofenamic acid. *Eur. J. Pharmacol.*, 378, 349-56.

In: Potassium Channels
Editor: Danielle S. Fonseca
ISBN: 978-1-61324-880-5

Chapter II

The Contribution of Lymphocyte Potassium Channels to the Perinatal Regulation of the Immune Response in Mother and Newborn

Gergely Toldi[1] and Barna Vásárhelyi[1,2]
[1]Research Group of Pediatrics and Nephrology,
Semmelweis University and Hungarian Academy of Sciences,
Budapest, Hungary
[2]Department of Laboratory Medicine, Semmelweis University,
Budapest, Hungary

Abstract

Voltage-gated Kv1.3 potassium channels and calcium-dependent IKCa1 potassium channels play an important role in the regulation of lymphocyte activation. Upon antigen presentation, they maintain the electrochemical driving force for sustained calcium influx that regulates cytokine production and further components of an adequate immune response.

Maternal and neonatal immune functions are characterized by distinct alterations in the perinatal period compared with the adult, non-pregnant immune status. These physiological characteristics play an important role in human reproduction and early postnatal development. At the same time, pathologic alterations of the immune response may occur both in the maternal and neonatal immune system, resulting in specific diseases. Recent studies demonstrated that lymphocyte potassium channels may be important elements in the development of the physiological and pathologic alterations of the immune system in human pregnancy and the neonatal period.

In healthy pregnancy, the maternal immune system acquires tolerance in order to protect the developing fetus from harmful immunological reactions. If this tolerance is impaired, an uncontrolled maternal immune response may arise, contributing to the development of a pregnancy specific syndrome, preeclampsia. There is a characteristic pattern of lymphocyte calcium influx and activation in healthy pregnancy influenced by potassium channels. This pattern is missing in preeclampsia, where the above properties are rather comparable to the non-pregnant state.

Decreased functionality of neonatal lymphocytes is a widely recognized experimental and clinical phenomenon. Cytokine production in activated T lymphocytes of the term neonate is reduced compared to adults. A possible contributing factor to this reduction might be the impairment of mechanisms regulating short-term activation of lymphocytes compared to adults. Kv1.3 and IKCa1 lymphocyte potassium channels are essential components of these mechanisms. Recent results indicate that characteristics of short-term activation of major neonatal lymphocyte subsets are indeed altered compared to adults. These findings improve the understanding of the mechanisms that prevent neonatal lymphocytes from adequate activation upon activating stimuli and, hence, exert a lower intensity of immune response. They show that the functional impairment of lymphocyte potassium channels may be of importance in those mechanisms.

In this chapter, we review recent knowledge on the contribution of lymphocyte potassium channels to perinatal immunity both from a maternal and a neonatal perspective. Furthermore, we provide data on the role of these channels in shaping the kinetics of calcium influx during lymphocyte activation in pregnancy and in neonates.

Introduction

The process of lymphocyte activation is essential for the development of adequate immune responses, and is closely linked to alterations of the cytoplasmic free calcium concentration ($[Ca^{2+}]_{cyt}$). The different antigens evoking an immunological reaction are presented to T lymphocytes by antigen presenting cells. These cells attach to the T cell receptor/CD3 (TCR/CD3) complex of T lymphocytes. The interaction between the presented antigen and the TCR/CD3 complex leads to a transient, biphasic increase of $[Ca^{2+}]_{cyt}$ through the activation of enzymes that are elements of transmembrane signaling pathways. This elevation is indispensable for the subsequent steps of T lymphocyte activation, such as the transcription of cytokines [1].

The calcium signal converges to the activation of nuclear factor of activated T cells (NF-AT). This is mediated by calcineurin which dephosporylates NF-AT, thus enabling it to accumulate in the nucleus and bind to the promoter region of the interleukin-2 (IL-2) gene. Once this gene is activated, T lymphocytes are able to proliferate even in the absence of an antigen. One of the most important enzymes activated by the antigen-launched signal transduction is phospholipase C-gamma (PLC-gamma). This enzyme cleaves phosphatidylinositol 4,5-bisphosphate (PIP_2) in the cell membrane to produce diacylglycerol (DAG) and inositol 1,4,5-trisphosphate (IP_3). The first phase of the biphasic calcium signal is directly tied to the generation of IP_3, since calcium is released from the endoplasmic reticulum (ER) through the binding of IP_3 to its designated receptor [2] (Figure 1).

The second phase of the increase of $[Ca^{2+}]_{cyt}$ originates from the activation of calcium release activated calcium (CRAC) channels in the cell membrane. CRAC belongs to the group of store operated ion channels, as it is activated by the discharge of intracellular calcium stores. The inward calcium current is primarily determined by the electrochemical driving force for calcium. In order to maintain the electrochemical potential gradient ($\Delta\ \Psi$) between the intracellular and extracellular space for further calcium entry, depolarizing calcium influx needs to be counterbalanced by the efflux of cations, predominantly potassium. Thus the relation between the ion currents through CRAC channels and the function of potassium channels makes the proliferation and activation of lymphocytes sensitive to pharmacological intervention regarding potassium channels, and provides an opportunity for specific immunomodulation [3].

There are two major types of potassium channels in T lymphocytes: the voltage-gated Kv1.3 and the calcium-activated IKCa1 channels. These channels have distinct functional properties.

Basic structural elements of Kv1.3 channels are similar to those of other voltage-gated potassium channels. These channels are composed of four identical, pore-forming subunits [4]. Each subunit consists of six transmembrane alpha helices connected by intra- and extracellular loops.

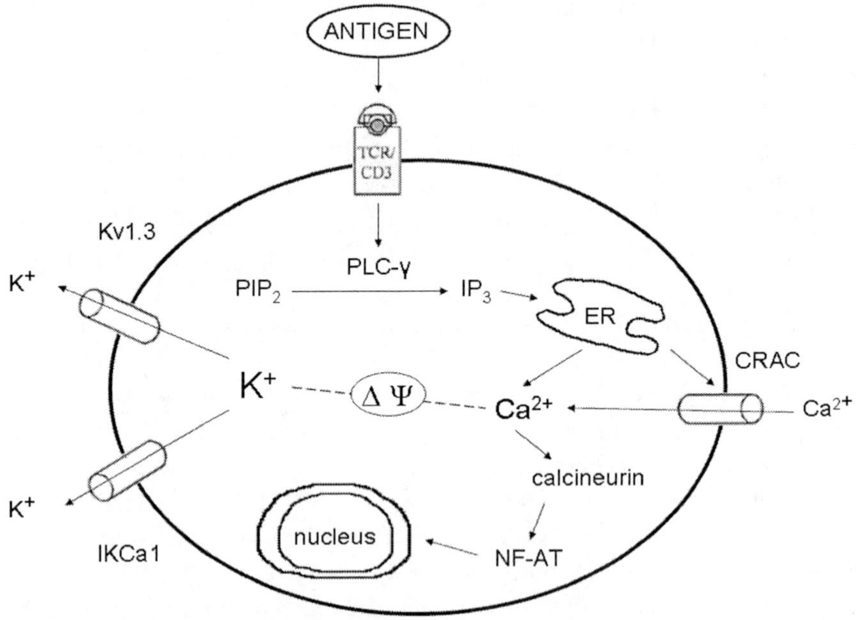

Figure 1. Simplified scheme of the role of Kv1.3 and IKCa1 lymphocyte potassium channels in the regulation of cytoplasmic free calcium concentration. TCR/CD3 – T-cell receptor/CD3 complex, PLC-γ – phospholipase C-gamma, PIP_2 – phosphatidylinositol 4,5-bisphosphate, IP_3 – inositol 1,4,5-trisphosphate, ER – endoplasmic reticulum, CRAC – calcium release activated calcium channel, Δ Ψ – electrochemical potential gradient.

The extracellular loops between the 5th and 6th transmembrane helices and segments of the 6th helix from each subunit form the pore through which potassium ions leave the cell. The chief structural elements of the conduction pore are the selectivity filter, followed by a water-filled cavity and the activation gate. Carbonyl oxygens on amino acids of the selectivity filter substitute water to guide potassium ions through the narrowest part of the channel [5]. The 4th alpha helix contains positively charged amino acid side chains, and is regarded as the voltage sensor of the channel. The activation threshold of Kv1.3 channels is between -50 mV and -60 mV [6].

The anticipated structure of IKCa1 channels is similar to that of voltage-gated potassium channels. It is composed of four identical, non-covalently linked subunits, each consisting of six transmembrane alpha helices. IKCa1 channels are activated by the rise of $[Ca^{2+}]_{cyt}$ above 100 nM [7]. Therefore, these channels do not function at the resting state of the cells, but are promptly activated during transmembrane signaling resulting in the elevation of

$[Ca^{2+}]_{cyt}$. Previous findings suggest that the calcium sensor of IKCa1 channels is calmodulin [8]. Interestingly, calcium is not needed for the association of calmodulin with IKCa1, indicating that calmodulin might be permanently linked to the channel. IKCa1 channels have an important role in the positive feedback regulation of the calcium signal in T lymphocytes. When $[Ca^{2+}]_{cyt}$ is elevated due to calcium influx through the CRAC channels, the increased potassium conductance of the membrane enables further calcium entry by maintaining the electrochemical potential gradient between the intra- and extracellular spaces.

Data obtained in previous investigations suggest that the contribution of Kv1.3 and IKCa1 channels to lymphocyte activation is different according to T lymphocyte subtypes. The activation of a T lymphocyte subset, the CCR7- CD45RA- effector memory T cells (TEM), primarily depends on Kv1.3 channels, while IKCa1 channels have a major role in the activation of CCR7+ CD45RA+ naïve and CCR7+ CD45RA- central memory T cells (TCM) [9]. Terminally differentiated TEM cells play a pivotal role in the pathogenesis of autoimmune disease. Wulff et al. suggest that disease causing TEM cells are able to home to inflamed tissues and exhibit immediate effector function and cause direct cell destruction during the autoimmune process. They reinforced that the characteristic potassium channel phenotype of TEM cells in multiple sclerosis is Kv1.3high IKCa1low, contrasting naïve and TCM cells, which exhibit a Kv1.3low IKCa1high channel phenotype [10]. Therefore the therapeutic relevance of specific Kv1.3 channel inhibitors is of outstanding interest, as they may offer the possibility for selective modulation of pathogenic TEM cells, while naïve and TCM cells (needed for physiological immune responses) would escape the inhibition through the upregulation of IKCa1 channel expression. Beeton et al. demonstrated that the symptoms of experimental autoimmune encephalitis, a murine model of multiple sclerosis, significantly improved after treatment with selective Kv1.3 inhibitors [11].

Peptide blockers of the Kv1.3 channel were isolated mostly from the venom of scorpions. These peptides bind into the outer vestibule of the channels and occlude the ion conduction pore [12]. They form multi-point contacts with several channel residues, which results in high affinity binding. One of the most potent blockers of the Kv1.3 channel is margatoxin (MGTX, Kd 50 pM), isolated from the venom of a scorpion species, Centruroides margaritatus [13].

The most effective inhibitor of the IKCa1 channel was found to be TRAM-34 (Kd 20-25 nM), a pyrazole-substituted triarylmethane (TRAM) compound [14]. This small molecule blocker binds to a site of the channel accessible from the intracellular side, blocking the intracellular end of the selectivity filter. Therefore, these molecules first need to cross the cell membrane. In contrast to peptide toxins, these blockers have a smaller number of contact areas with the channel, resulting in a lower affinity binding compared with peptide inhibitors.

Nonetheless, the use of potassium channel inhibitors for the treatment of human T cell mediated disorders is hampered by the lack of knowledge on the functional significance of potassium channels in the process of lymphocyte activation in major human T cell subpopulations isolated from healthy and diseased individuals. In our recent investigations [15, 16], we studied the role of the Kv1.3 and IKCa1 lymphocyte potassium channels in the process of lymphocyte activation using a flow cytometry method in two physiological states with distinct immunological features (i.e. in samples taken from pregnant women and neonates). Both in pregnancy and in the neonatal period, the immune response and the kinetics of lymphocyte activation are altered compared with the adult, non-pregnant state. Lymphocyte potassium channels play a pivotal role in the regulation of the immune response; however, they show altered functionality in pregnancy and in the neonate.

Flow-Cytometry for Monitoring Lymphocyte Activation

Until recently, principally single-cell techniques were applied to study the process of lymphocyte activation, and no reliable high-throughput method was available to investigate lymphocyte activation kinetics in more than one lymphocyte subsets simultaneously. The use of single-cell techniques are limited by the fact that they are not suitable to describe this process in a complex cellular milieu that contains different types of interacting immune cells. For this purpose, we developed a novel approach that enabled us to monitor lymphocyte activation simultaneously in different lymphocyte subtypes. It is increasingly being accepted that this novel method provides new insight into the functionality of potassium channels in major human lymphocyte subsets. For our measurements, we collected peripheral blood samples from adult subjects. In case of neonates, cord blood samples were collected. Informed consent was obtained from all participating subjects, or, in case of neonates, from the parents of subjects, and our studies were reviewed and approved by an independent ethical committee of the institution.

Peripheral blood mononuclear cells (PBMCs) were separated by a standard density gradient centrifugation as previously described [15, 16]. PBMCs were incubated with conjugated anti-human monoclonal antibodies for surface marker staining and were loaded with calcium sensitive Fluo-3 and Fura Red dyes to monitor alterations of $[Ca^{2+}]_{cyt}$ as described before.

PBMCs were divided into three vials. One vial was treated with MGTX (60 nM), a selective blocker of the Kv1.3 channel. Another vial was treated with TRAM-34 (60 nM), a specific inhibitor of the IKCa1 channel. The third vial was used as control. Measurements were initiated directly after the addition of 20 μg of phytohemagglutinin (PHA) as an unspecific activating stimulus. Cell fluorescence data were measured and recorded for 10 minutes in a kinetic manner on flow cytometer. Data acquired from the measurements were evaluated with our specific software. The core of this software is an improved version of an algorithm [17] based on the calculation of a double-logistic function for each recording. This function is used to describe measurements that have an increasing and a decreasing intensity as time passes. The software also calculates parameter values describing each function, such as the Maximum value (Max), the Time to reach maximum value (t_{max}), and the Area Under the Curve (AUC) (Figure 2).

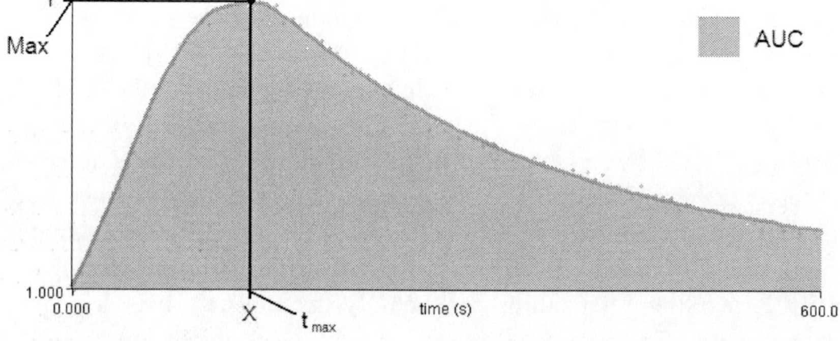

Figure 2. Parameter values calculated for each flow cytometry recording. One unit of the Area Under the Curve value is defined as one relative intensity value in one second, where relative intensity values are the rate of actual intensity values divided by intensity values at zero second. Relative parameter value represents the relative fluorescence of calcium binding dyes (ratio of Fluo-3 and Fura Red). Max – Maximum value (Y), t_{max} – Time to reach maximum value (X), AUC – Area Under the Curve.

These parameters represent different characteristics of lymphocyte calcium influx kinetics. The Maximum value represents the peak value of the calcium influx curve upon lymphocyte activation, thus it reflects the maximal amount of $[Ca^{2+}]_{cyt}$ in the course of activation. The Time to reach maximum value describes how soon the peak value of the calcium influx curve is reached. The Area Under the Curve describes the full amount of $[Ca^{2+}]_{cyt}$ during the whole period of lymphocyte activation recorded and thus the magnitude of the elicited calcium response in general. AUC values correspond to the sum of $[Ca^{2+}]_{cyt}$ increase, which further corresponds to the level of lymphocyte activation [18].

Lymphocyte Activation in Pregnancy

To ensure a healthy pregnancy, the maternal immune system needs to adapt to the novel physiological needs raised by the developing fetus. Since the conceptus is half of foreign origins, presenting paternal antigens, it is considered a semi-allograft to maternal immunity. Therefore, an immune tolerance must develop to avoid immunological rejection of the fetus. In the normal course of pregnancy, the mother extends her 'definition of self' for 40 weeks on the foreign antigens of the fetus. The impairment of this tolerance and the development of an abnormal immune response play a major role in adverse pregnancy outcomes, including preeclampsia [19]. This disorder is characterized by hypertension, proteinuria, edema and endothelial dysfunction usually evolving in the third trimester of pregnancy; however, it can also occur earlier. Although preeclampsia is quite common (i.e. it affects about 5-8% of all pregnancies globally), its clear cause and the mechanisms leading to immune dysfunction remain to be elucidated. Preeclampsia is estimated to be responsible for about 70 000 maternal deaths each year worldwide [20]. HELLP syndrome (standing for hemolysis, elevated liver enzymes, low platelet count) and eclampsia are other manifestations of the same disorder. Although these conditions are generally coupled with a number of other symptoms (including headaches, abdominal pain, nausea, vomiting, changes in vision, shortness of breath, anxiety, mental confusion, seizures), these manifestations are not necessarily more serious than preeclampsia. Besides a maternal systemic inflammatory response, signs of systemic vasoconstriction (which might be a consequence of inflammation) may also be observed in the mother in these pregnancy-associated disorders [21].

Several aspects of the development of the pregnancy-specific immune tolerance have been described recently. Initially, the contact between maternal and fetal cells is taking place on a local level and is restricted to the decidua, but during the second trimester of pregnancy, it is extended to the entire body of the mother. Both the innate and adaptive, as well as the humoral and cellular arms of immunity are involved in these events.

For many years, it has been hypothesized that normal pregnancy induces a shift from Th1 immunity towards Th2 immunity. However, it has been demonstrated recently that the levels of particular Th1 cytokines are raised, instead of lowered in normal pregnancy compared with the non-pregnant state. Current findings indicate that gravidity is both a pro-inflammatory and an anti-inflammatory condition, depending on the stage of gestation. Grossly, pregnancy has three distinct immunological phases. The events of implantation and the first trimester require a strong inflammatory response to ensure the adequate remodeling of the uterine epithelium and the removal of cellular debris following the implantation. Thus, the first trimester of pregnancy is hallmarked by pro-inflammatory events. The second immunological phase of gravidity consists of the second and third trimesters. This is the period of fetal growth and development, when an anti-inflammatory state is established. Finally, delivery represents the shortest, third immunological phase of pregnancy, when pro-inflammatory events dominate again and promote uterinal contractions to deliver the fetus and the placenta [22, 23].

Under pathologic conditions with insufficient immune tolerance such as in preeclampsia, the anti-inflammatory state during the second and third trimesters does not adequately develop. Preeclampsia is characterized by a cytokine-mediated excessive maternal inflammatory response. An important feature of this disorder is the absence of Th2 skewness and thus the predominance of pro-inflammatory cytokines, as shown by a number of investigations [24-26].

Previous studies demonstrated that not only the prevalence of T lymphocyte subsets, but also the functionality of these cells is altered in pregnancy. For instance, calcium handling of T lymphocytes and, therefore, lymphocyte activation kinetics differ in healthy pregnancy compared with the non-pregnant state. Indeed, previous reports indicate a sustained increase of basal intracellular calcium level in lymphocytes of healthy pregnant and preeclamptic women compared to non-pregnant women, with the highest levels in preeclampsia [27-29]. We hypothesized that the elements of calcium handling of activated T lymphocytes, including lymphocyte potassium channels may be affected in healthy pregnancy and preeclampsia compared to the non-pregnant state. Our group characterized the activation-elicited calcium influx in major T lymphocyte subsets (i.e. Th1, Th2, CD4 and CD8 cells) in healthy pregnant, preeclamptic and non-pregnant women and tested its alteration upon the inhibition of Kv1.3 and IKCa1 lymphocyte potassium channels.

The Contribution of Potassium Channels to Lymphocyte Activation in Pregnancy

Our results indicate that calcium influx kinetics in activated T lymphocytes markedly differs in healthy pregnancy compared to the non-pregnant state: AUC values of the calcium response are lower in healthy pregnancy in the Th1, CD4 and CD8 lymphocyte subsets [15]. Based on this observation, it is reasonable to assume that the physiological immune tolerance towards fetal antigens in pregnancy is partly attributed to a lower calcium response. This hypothesis is further supported by the particular role of the impaired function of Th1 and CD8 lymphocyte subsets in maternal immune tolerance [30]. On contrary to Th1 cells, the activation induced calcium response of the Th2 subset is not decreased compared with the non-pregnant state. The decreased activation of the Th1 subset (reflected by low AUC values in our study) and the lack of decrease in Th2 cells may partly be responsible for the well established Th2 skewness in healthy pregnancy [24-26].

Unlike in healthy pregnancy, we could not detect a difference in the AUC values of calcium influx kinetics of Th1 and CD8 cells in preeclampsia compared to non-pregnant women. The absence of calcium influx characteristics specific for healthy pregnancy suggests that this element may associate with the impaired maternal immune tolerance present in preeclampsia, since the calcium influx kinetics is comparable to that seen in non-pregnant samples [30]. Indeed, the maintained activation properties of Th1 lymphocytes in preeclamptic patients may contribute to the lack of Th2 dominance associated with normal pregnancy. Similarly to the Th1 subset, CD8 cells in preeclampsia are also characterized by the lack of suppressed activation kinetics [31]. Thus the decrease of cytotoxic activity observed in healthy pregnancy is not present in preeclampsia [32].

Interestingly, t_{max} values were decreased in Th2 and CD4 cells in preeclampsia compared with healthy pregnancy. This finding may indicate an increased reactivity of lymphocytes in preeclampsia, possibly reflecting an elevated responsiveness of T lymphocytes due to the ongoing maternal systemic inflammation.

Since Kv1.3 and IKCa1 potassium channels significantly influence the calcium response elicited upon lymphocyte activation, we tested their expression and function in healthy pregnancy and preeclampsia. According to comparable fluorescence values of the samples stained with specific antibodies against Kv1.3 channels, their expression is not altered in any of the investigated T lymphocyte subsets. Therefore, the differences detected between calcium influx of non-pregnant, healthy pregnancy and preeclamptic lymphocytes upon treatment with specific inhibitors of the potassium channels is probably due to the altered function, and not to the altered expression of these channels.

Our results suggest that the overall lymphocyte population and particularly the CD4 subset are sensitive to MGTX and TRAM inhibition in each investigated group of subjects, indicating that both Kv1.3 and IKCa1 channels play an important regulatory role in calcium influx in these lymphocyte populations. This is reflected by the decrease of the AUC and Max values compared with the respective samples where no inhibitors were applied. However, the sensitivity of calcium influx measured in other lymphocyte subsets shows clear variability. It is of particular interest that calcium influx of Th2 lymphocytes in healthy pregnancy was insensitive to potassium channel inhibition, while calcium influx decreased significantly in non-pregnant samples upon treatment with the specific channel blockers. Of note, Th2 lymphocytes in preeclampsia presented with non-pregnant-like characteristics, and were also sensitive to MGTX and TRAM treatment. Since the regulatory function of Kv1.3 and IKCa1 channels on calcium influx appears to be limited in healthy pregnant samples (as the inhibition of these channels did not result in a decrease of the AUC and Max values), it is tempting to speculate that this may be an element contributing to the Th2 shift present in healthy pregnancy, but absent in preeclampsia. This hypothesis may be supported by reports suggesting that the shape of calcium influx influenced by potassium channel functions may determine the cytokine production profile of helper T lymphocytes [33].

Interestingly, other differences were also observed between healthy pregnancy and preeclampsia. While calcium influx in CD8 and Th1 lymphocytes was resistant to potassium channel inhibition in preeclamptic, that of healthy pregnant lymphocytes was sensitive. Similarly to Th2 cells, while it is unclear whether the resistance of Th1 lymphocytes to potassium channel inhibition is reflected in their function, the insensitivity of the Th1 subset to the inhibition of regulatory lymphocyte potassium channels in preeclampsia may be linked to the Th1 skewness [34].

t_{max} values, representing the kinetics of lymphocyte activation, were not altered upon inhibitory treatment in any of the investigated subsets. This finding reflects that it is rather the amount of calcium entering the cells, and not the rate of entry that is influenced by potassium channel inhibition in pregnancy.

Lymphocyte Activation in the Neonate

The immune response in the neonate is characterized by a higher level of plasticity when compared with adults.

In addition to the development of protective immune responses against infectious agents, the newborn is also presented with a wide range of novel environmental antigens that do not require the onset of an immunological reaction. Neonatal life is the period when tolerance towards most of these antigens needs to be established. It is also in this period of life that tolerance to peripheral tissue antigens of the self must develop to avoid autoimmune reactions. Therefore, the function of T lymphocytes must differ from that seen in adults to be able to adapt to the specific challenges raised in early life.

Inevitably, decreased functionality of neonatal T cells is a widely recognized experimental and clinical phenomenon. Reduced functioning is well characterized by a lower level of cytokine production compared to adult T cells. It is in particular the level of pro-inflammatory cytokines (such as IFN-gamma) that are lower in the neonate [35, 36]. Several factors might be responsible for the decreased cytokine expression compared to adult lymphocytes.

Previous studies suggest that neonatal T cells respond weakly to physiological stimuli, i.e. activation by APCs via the TCR/CD3 complex. Nevertheless, they are capable of adult-level IL-2 production and proliferation when the TCR/CD3 complex is bypassed. Therefore, since cytokine production can be boosted to adult levels by TCR-independent stimulation, there is probably no intrinsic deficiency in the capacity of neonatal T helper cells. Furthermore, neonatal T cells can produce adult levels of cytokines in response to TCR-dependent stimulation by increasing the number of costimulatory signals that promote Th1 cell function.

Finally, in the presence of adult APCs, the proliferation of neonatal T cells is also increased to adult levels. Therefore, neonatal T cells appear to have a greater requirement for costimulatory signals than adult T cells do. When costimulation is sufficiently provided, neonatal T cells achieve adult-level function.

This is, at least in part, due to the fact that the number of TCR/CD3 complexes is reduced on neonatal T cells, compared with that on adult T cells, probably resulting in a lower level of signal transduction via the TCR/CD3 complex. Additionally, cellular adhesion molecules, such as leukocyte functional antigen-1 (LFA-1) and CD2 are expressed at lower densities on neonatal T cells, further contributing to decreased activation [37].

In neonates, poor cytokine production can partially be attributed to the lack of previous antigen exposure and hence, the naïvity of most cells. The majority of these cells are naïve (CD45RA+) lymphocytes in contrast to adults, where effector (CD45RO+) cells dominate [36].

Furthermore, an interesting study by Hassan et al. has suggested that the CD45RA+ populations within cord and adult peripheral blood differ in their cytokine production since purified CD4+ CD45RA+ cells derived from cord blood do not produce IL-2 upon stimulation, whereas purified adult CD4+ CD45RA+ cells do [38]. Thus, cord blood CD45RA+ cells may be more naïve than adult CD45RA+ cells.

Another possible factor might be the impairment of mechanisms regulating short-term activation of lymphocytes, such as Kv1.3 and IKCa1 lymphocyte potassium channels, compared to adults. As described above, specific inhibition of these channels results in a diminished calcium influx into lymphocytes and a lower level of lymphocyte activation.

As a result, the cytokine expression profile of lymphocytes decreases. Indeed, early observations support that $[Ca^{2+}]_{cyt}$ of lymphocytes (that is influenced by potassium channels) is strongly associated with IFN-gamma production [39].

In our investigations, we hypothesized that short-term T lymphocyte activation properties are different in neonates compared to adults. We aimed to characterize the calcium influx kinetics upon activation in major T lymphocyte subsets (i.e. Th1, Th2, CD4 and CD8 cells) in the neonate, and its sensitivity to the specific inhibition of Kv1.3 and IKCa1 lymphocyte potassium channels.

The Contribution of Potassium Channels to Lymphocyte Activation in the Neonate

Our findings revealed that calcium influx following PHA activation of T lymphocytes markedly differs in the neonate from that in the adult. The lower AUC and Max values suggest that short-term activation and associated calcium influx kinetics are decreased in neonatal lymphocytes. These results are in line with the widely known fact that newborns mount lower immune responses to distinct stimuli. Previous findings also indicate the decreased reactivity of neonatal (cord blood) T lymphocytes compared with adults [35, 37]. Neonatal lymphocytes mount poorer responses compared with adults, and require more costimulatory signals besides TCR-dependent stimulation to achieve adult level activation and cytokine production [37].

t_{max} values were also decreased in the Th1 and Th2 subsets in neonates compared with adults. The level of decrease was greater in Th1 cells, indicating that this subset is activated more rapidly upon PHA stimulation. Therefore, it seems that in neonates, T helper lymphocytes are activated sooner upon adequate stimuli when a pro-inflammatory respond is needed compared with effects that require an anti-inflammatory response. This suggests that establishing a pro-inflammatory response in achieved more easily in the neonatal period than acquiring an anti-inflammatory state. This assumption is probably not supported by data obtained from murine neonates, where Th2 function dominates over Th1 function [37]. However it is not known whether this also implies a more rapid response given by this subset in mice in comparison with Th1 cells.

Upon treatment of lymphocytes with selective inhibitors of the Kv1.3 and IKCa1 channels (MGTX and TRAM, respectively), calcium influx decreases in most investigated lymphocyte subsets isolated from adults. This decrease is represented by the reduction of the AUC and Max values. However, with the exception of the CD8 subset, such a reduction was not demonstrated in neonatal lymphocytes.

The finding that neonatal lymphocytes are less sensitive to the specific inhibition of potassium channels compared to adults may be due to altered functionality or a lower expression of these channels. Therefore, we measured the expression of Kv1.3 potassium channels on the investigated lymphocytes. Instead of lower values, we found increased expression of Kv1.3 channels on neonatal CD4, CD8 and Th2 lymphocytes compared to adults. Thus the option that the decreased sensitivity of lymphocytes to potassium channel inhibition is due to a lower channel expression should be excluded. Based on the lower sensitivity of lymphocyte calcium influx to inhibition at higher channel expression values, it is reasonable to postulate that neonatal Kv1.3 channels are functionally altered. Similar observations have been published regarding the expression and dysfunction of Ca^{2+}-ATPase [40] and Na^{+}/K^{+}-ATPase [41] in neonatal erythrocytes. These enzymes are more abundant in preterm than in term neonates, and their numbers decrease during the perinatal development. However, enzyme activities were measured to be similar in both preterm and full term neonates. The same activity at higher enzyme molecule numbers might indicate a potential immaturity of the enzyme in the preterm infant.

Of note, we observed tendencies similar to Kv1.3 channels by specific inhibition of IKCa1 channels. The lack of available antibodies prevented us from obtaining data on IKCa1 expression. However, indirect evidence does refer to the altered functionality of IKCa1 channels in newborns. We measured significantly increased basal median fluorescence values of calcium binding dyes that may indicate higher $[Ca^{2+}]_{cyt}$ levels in neonatal lymphocytes than in adult ones. An increased basal $[Ca^{2+}]_{cyt}$ level is expected to increase cell excitability, cytokine production capacity [1] and also IKCa1 channel function [7]. The lower excitability and decreased sensitivity to IKCa1 channel inhibition coupled with the contradictory finding of increased basal $[Ca^{2+}]_{cyt}$ in neonatal lymphocytes suggest that IKCa1 channels are altered in function compared to their function in adults.

Signs of altered sensitivity to potassium channel inhibition were present in all major lymphocyte subsets investigated (i.e. Th1, Th2, CD4 and CD8 cells). The only subset in newborns in which the short-term activation was inhibited significantly by specific blockers of both Kv1.3 and IKCa1 channels was CD8 lymphocytes. However, even in this case, the level of inhibition did not reach the extent described in adults in spite of high Kv1.3 expression values. This suggests that our observations are generally characteristic of all lymphocyte subpopulations studied.

Evaluating our results it should also be taken into consideration that the majority of lymphocytes are naïve (CD45RA+) cells in the neonate, and effector (CD45RO+) lymphocytes have a lower prevalence compared to adults. Thus the difference in the prevalence of naïve and effector lymphocytes might be a component of the differences we describe in the calcium influx characteristics of neonates compared to adults. However, since Vig et al. demonstrated that naïve T lymphocyte populations show no defect in calcium influx in the presence of physiological concentrations of extracellular calcium [42], this option is less likely.

Conclusion

In our investigations, we reported on calcium influx kinetics upon lymphocyte activation and the functionality of Kv1.3 and IKCa1 lymphocyte potassium channels in healthy pregnancy, preeclampsia, and in neonates. Our results indicate marked differences of calcium influx kinetics (Figure 3) and sensitivity to lymphocyte potassium channel inhibition in major lymphocyte subsets between non-pregnant, healthy pregnant and preeclamptic lymphocytes. It is of interest that these properties in preeclampsia are more comparable to the non-pregnant state than to healthy pregnancy. These findings suggest that there is a characteristic pattern of calcium influx and its sensitivity to potassium channel inhibition in healthy pregnant women that is missing in preeclamptic patients. This raises the notion that lymphocyte calcium handling upon activation may have a role in the characteristic immune status of healthy pregnancy.

Furthermore, our results may partly explain why neonatal lymphocytes are less responsive to activating stimuli and, hence, exert a lower intensity of immune response. We demonstrated that short-term activation and associated calcium influx kinetics are lower in neonatal lymphocytes compared to adults (Figure 3).

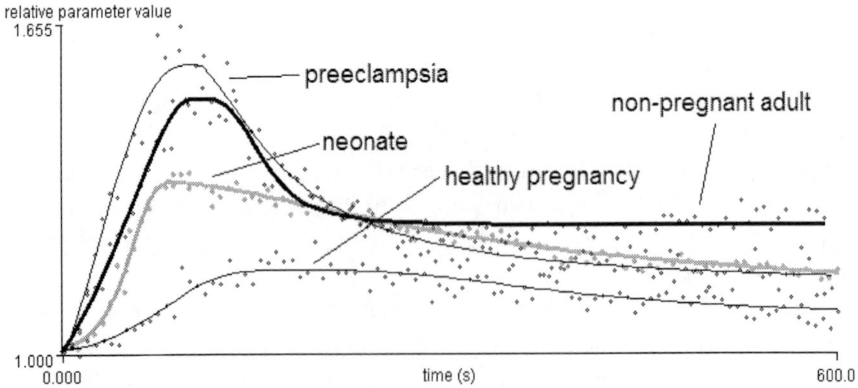

Figure 3. Calcium influx kinetics of the Th1 lymphocyte subset (without inhibitor treatment) isolated from non-pregnant adult, healthy pregnant, preeclamptic, and newborn subjects. Relative parameter value represents the relative fluorescence of calcium binding dyes (ratio of Fluo-3 and Fura Red). Calcium influx is lower in healthy pregnancy than in the non-pregnant state or preeclampsia. Calcium influx is also decreased in neonates compared with (non-pregnant) adults.

This phenomenon is associated with the altered function of lymphocyte potassium channels. Our results improve the understanding of the mechanisms that prevent neonatal T cells from adequate activation upon activating stimuli, and partially elucidate previous experimental data indicating that a greater amount of stimulation is needed in neonatal lymphocytes compared with adults to achieve a similar immune response. The functional impairment of lymphocyte potassium channels may be of importance in those mechanisms.

Potassium channels are important contributors to the control of T lymphocyte activation. Their altered functionality plays a significant role in the development of specific differences of the immune response in pregnancy and in the neonate. Further characterization of the influence of lymphocyte potassium channels on calcium influx kinetics in T cells will broaden our knowledge on the specific immunological processes in these conditions. Additionally, the understanding of Kv1.3 and IKCa1 channel function in autoimmune disorders might advance the design of novel therapeutic strategies targeting the modulation of the excessive immune response.

References

[1] Crabtree, GR; Clipstone NA. Signal transmission between the plasma membrane and nucleus of T lymphocytes. *Annu. Rev. Biochem*, 1994, 63, 1045-1083.

[2] Lewis, RS. Calcium signaling mechanisms in T lymphocytes. *Annu. Rev. Immunol.,* 2001, 19, 497-521.

[3] Zweifach, A; Lewis, RS. Mitogen-regulated Ca2+ current of T lymphocytes is activated by depletion of intracellular Ca2+ stores. *Proc. Natl. Acad. Sci. USA,* 1993, 90, 6295-6299.

[4] Vicente, R; Escalada, A; Soler, C; Grande, M; Celada, A; Tamkun, MM; et al. Pattern of Kv beta subunit expression in macrophages depends upon proliferation and the mode of activation. *J. Immunol.*, 2005, 174, 4736-4744.

[5] Doyle, DA; Morais, CJ; Pfuetzner, RA; Kuo, A; Gulbis, JM; Cohen, SL; et al. The structure of the potassium channel: molecular basis of K+ conduction and selectivity. *Science*, 1998, 280, 69-77.

[6] Matteson, DR; Deutsch, C. K channels in T lymphocytes: a patch clamp study using monoclonal antibody adhesion. *Nature*, 1984, 307, 468-471.

[7] Grissmer, S; Nguyen, AN; Cahalan, MD. Calcium activated potassium channels in resting and activated human T lymphocytes. Expression

levels, calcium dependence, ion selectivity, and pharmacology. *J. Gen. Physiol.*, 1993, 102, 601-630.

[8] Fanger, CM; Ghanshani, S; Logsdon, NJ; Rauer, H; Kalman, K; Zhou, J; et al. Calmodulin mediates calciumdependent activation of the intermediate conductance KCa channel, IKCa1. *J. Biol. Chem.*, 1999, 274, 5746-5754.

[9] Wulff, H; Beeton, C; Chandy, KG. Potassium channels as therapeutic targets for autoimmune disorders. *Curr. Opin. Drug Discov. Devel.*, 2003, 6, 640-647.

[10] Wulff, H; Calabresi, PA; Allie, R; Yun, S; Pennington, M; Beeton, C; et al. The voltage-gated Kv1.3 K+ channel in effector memory T cells as new target for MS. *J. Clin. Invest,* 2003, 111, 1703-1713.

[11] Beeton, C; Wulff, H; Standifer, NE; Azam, P; Mullen, KM; Pennington, MW; et al. Kv1.3 channels are a therapeutic target for T cell-mediated autoimmune diseases. *Proc. Natl. Acad. Sci. USA,* 2006, 103, 17414-17419.

[12] Goldstein, SA; Miller, C. Mechanism of Charybdotoxin block of a voltage-gated K+ channel. *Biophys. J.,* 1993, 65, 1613-1619.

[13] Garcia-Calvo, M; Leonard, RJ; Novick, J; Stevens, SP; Schmalhofer, W; Kaczorowski, GJ; et al. Purification, characterization, and biosynthesis of margatoxin, a component of Centruroides margaritatus venom that selectively inhibits voltage-dependent potassium channels. *J. Biol. Chem.*, 1993, 268, 18866-18874.

[14] Wulff, H; Gutman, GA; Cahalan, MD; Chandy, KG. Delineation of the clotrimazole/TRAM-34 binding site on the intermediate conductance calcium-activated potassium channel, IKCa1. *J. Biol. Chem.*, 2001, 276, 32040-32045.

[15] Toldi, G; Stenczer, B; Treszl, A; Kollar, S; Molvarec, A; Tulassay, T; et al. Lymphocyte calcium influx characteristics and their modulation by Kv1.3 and IKCa1 channel inhibitors in healthy pregnancy and preeclampsia. *Am. J. Reprod. Immunol.*, 2010, in press.

[16] Toldi, G; Treszl, A; Pongor, V; Gyarmati, B; Tulassay, T; Vasarhelyi, B. T-lymphocyte calcium influx characteristics and their modulation by Kv1.3 and IKCa1 channel inhibitors in the neonate. *Int. Immunol.*, 2010, 22, 769-774.

[17] Kaposi, AS; Veress, G; Vásárhelyi, B; Macardle, P; Bailey, S; Tulassay, T; et al. Cytometry-acquired calcium-flux data analysis in activated lymphocytes. *Cytometry A*, 2008, 73, 246-253.

[18] Panyi, G. Biophysical and pharmacological aspects of K+ channels in T lymphocytes. *Eur. Biophys. J.*, 2005 34, 515-529.

[19] Saito, S; Shiozaki, A; Nakashima, A; Sakai, M; Sasaki, Y. The role of the immune system in preeclampsia. *Mol. Aspects Med.,* 2007, 28, 192-209.

[20] Walker, JJ. Pre-eclampsia. *Lancet,* 2000, 356: 1260-1265.

[21] Baumwell, S; Karumanchi, SA. Pre-eclampsia: clinical manifestations and molecular mechanisms. *Nephron Clin. Pract.*, 2007, 106, c72-81.

[22] Mor, G; Cardenas, I. The immune system in pregnancy: a unique complexity. *Am. J. Reprod.Immunol.*, 2010, 63, 425-433.

[23] Challis, JR; Lockwood, CJ; Myatt, L; Norman, JE; Strauss, JF 3rd; Petraglia, F. Inflammation and pregnancy. *Reprod. Sci.*, 2009, 16, 206-215.

[24] Saito, S; Umekage, H; Sakamoto, Y; Sakai, M; Tanebe, K; Sasaki, Y; et al. Increased T-helper-1-type immunity and decreased T-helper-2-type immunity in patients with preeclampsia. *Am. J. Reprod. Immunol.*, 1999, 41, 297-306.

[25] Rein, DT; Schondorf, T; Gohring, UJ; Kurbacher, CM; Pinto, I; Breidenbach, M; et al. Cytokine expression in peripheral blood lymphocytes indicates a switch to T(HELPER) cells in patients with preeclampsia. *J. Reprod. Immunol.*, 2002, 54, 133-142.

[26] Darmochwal-Kolarz, D; Rolinski, J; Leszczynska-Goarzelak, B; Oleszczuk, J. The expressions of intracellular cytokines in the lymphocytes of preeclamptic patients. *Am. J. Reprod. Immunol.*, 2002, 48, 381-386.

[27] Hojo, M; Suthanthiran, M; Helseth, G; August, P. Lymphocyte intracellular free calcium concentration is increased in preeclampsia. *Am. J. Obstet Gynecol.*, 1999, 180, 209-214.

[28] von Dadelszen, P; Wilkins, T; Redman, CW. Maternal peripheral blood leukocytes in normal and pre-eclamptic pregnancies. *Br. J. Obstet. Gynaecol.,* 1999, 106, 576-581.

[29] Thway, TM; Shlykov, SG; Day, MC; Sanborn, BM; Gilstrap, LC 3rd; Xia, Y; et al. Antibodies from preeclamptic patients stimulate increased intracellular Ca2+ mobilization through angiotensin receptor activation. *Circulation,* 2004, 110, 1612-1619.

[30] Sargent, IL; Borzychowski, AM; Redman, CW. Immunoregulation in normal pregnancy and pre-eclampsia: an overview. *Reprod. Biomed. Online*, 2006, 13, 680-686.

[31] Darmochwal-Kolarz, D; Saito, S; Rolinski, J; Tabarkiewicz, J; Kolarz, B; Leszczynska-Gorzelak, B; et al. Activated T lymphocytes in pre-eclampsia. *Am. J. Reprod. Immunol.*, 2007, 58, 39-45.

[32] Malinowski, A; Szpakowski, M; Tchórzewski, H; Zeman, K; Pawlowicz, P; Wozniak, P. T lymphocyte subpopulations and lymphocyte proliferative activity in normal and pre-eclamptic pregnancy. *Eur. J. Obstet Gynecol. Reprod. Biol.*, 1994, 53, 27-31.

[33] Dolmetsch, RE; Xu, K; Lewis, RS. Calcium oscillations increase the efficiency and specificity of gene expression. *Nature,* 1998, 392, 933-936.

[34] Fanger, CM; Neben, AL; Cahalan, MD. Differential Ca2+ influx, KCa channel activity, and Ca2+ clearance distinguish Th1 and Th2 lymphocytes. *J. Immunol.*, 2008, 164, 1153-1160.

[35] Cohen, SB; Perez-Cruz, I; Fallen, P; Gluckman, E; Madrigal, JA. Analysis of the cytokine production by cord and adult blood. *Hum. Immunol,* 1999, 60, 331-336.

[36] García Vela, JA; Delgado, I; Bornstein, R; Alvarez, B; Auray, MC; Martin, I; et al. Comparative intracellular cytokine production by in vitro stimulated T lymphocytes from human umbilical cord blood (HUCB) and adult peripheral blood (APB). *Anal. Cell Pathol.*, 2000, 20, 93-98.

[37] Adkins, B. T-cell function in newborn mice and humans. *Immunol. Today,* 1999, 20, 330-335.

[38] Hassan, J; Reen, D. Cord blood CD4+ CD45RA+ T cells achieve a lower magnitude of activation when compared with their adult counterparts. *Immunology,* 1997, 90, 397-401.

[39] Kesson, AM; Bryson, YJ. Uptake of extracellular Ca2+ is a requirement for production of interferon-gamma by cord blood mononuclear cells. *J. Interferon Res.,* 1991, 11, 81-86.

[40] Kocsis, I; Vásárhelyi, B; Héninger, E; Vér, A; Tulassay, T. Expression and activity of the Ca(2+)-atpase enzyme in human neonatal erythrocytes. *Biol. Neonate*, 2001, 80, 215-218.

[41] Vasarhelyi, B; Tulassay, T; Ver, A; Dobos, M; Kocsis, I; Seri, I. Developmental changes in erythrocyte Na(+),K(+)-ATPase subunit abundance and enzyme activity in neonates. *Arch. Dis. Child Fetal Neonatal. Ed,* 2000, 83, F135-138.

[42] Vig, M; Kinet, JP. Calcium signaling in immune cells. *Nat. Immunol.*, 2009, 10, 21-27.

In: Potassium Channels
Editor: Danielle S. Fonseca

ISBN: 978-1-61324-880-5

Chapter III

Mitochondrial Potassium Channel: Features and Physiological Participation

R. Milan and F. Martinez*
Departamento de Bioquímica,
Facultad de Medicina Universidad Nacional Autónoma de México,
Ciudad de México, México

Cellular homeostasis is fundamental to maintain the cell alive and in this context ions play an important role. Ion movement is tightly controlled by a variety of channels that function as gateways into the cell. All channels are selective for the type of ion that can pass through its pore (Zhou 2010). The opening of these channels is tightly regulated by several stimuli, such as voltage, pH or the binding of specific molecules. Definite actions on the cellular metabolism are observed once the increase of ions takes place in the cytoplasm. Then, the movement of ions is related to cell signaling acting as second messengers.

Potassium channels are the most studied. In general, the main strategy to known their structural composition and functions has been made by mutations.

* Corresponding author at: Departamento de Bioquimica, Facultad de Medicina, UNAM, Apdo. Postal 70-159, 04510, Mexico City, Mexico. Tel: 52 55 5623 2168; fax: 52 55 5616 2419. E-mail address: fedem@bq.unam.mx

Similar experiments contributed to know the composition of channels as well as to identify the channel gate. More than 75 different mammalian genes coding for the potassium channels have been described. The channel diversity can be explained as the result of an alternative splicing during mRNA processing. There must be more functional channels than those codified in the genes, since different subunits of the same family can co-assemble to form diverse functional channels (Jenkinson 2006).

An important amount of studies on potassium channel have been made *in situ* with mRNA hybridization and immunohistochemistry. In this sense, electrophysiological and pharmacological experimental strategies have been useful giving strong basis to show that the channel is functional, found in the cellular membrane and co-assembled correctly (Olson et al. 2010).

The main criterion to classify K^+ channels is the number of crosses of the protein in the membrane; defined as transmembrane (TM) spanning regions. There are three main families of subunits that have two, four or six TM regions. All of these subunits have sequences containing the following five amino acids: threonine, valine, glycine, tyrosine and glycine. The sequences are highly conserved in all potassium channels (Doyle et al. 1998).

The potassium channels expressed in the cell plasma membrane have the function of maintaining a particular concentration of K^+ among different cellular compartments. For instance, cytoplasm has a potassium concentration of 120-140 mM and mitochondria of 160-180 mM, suggesting that the channels contribute importantly to the generation of chemical gradients. Other functions of potassium channels, such as the adenosine triphosphate (ATP)-sensitive K^+ (KATP) channel (that is a combination of an inward rectifier K^+ channel and an ATP-binding cassette protein) is to be a sensor of the metabolism, since the malfunctioning of this channel has been implicated in the development and progression of heart disease (Alekseev et al. 2010). K_{ATP} channels are abundantly expressed in the cardiovascular system, where they participate in the regulation of coronary blood flow under normal basal conditions (Brayden 2002) and in the pathophysiological ischemia/reperfusion cardiac injury. It has been observed that hearts with a short period of ischemia will be protected from a subsequent ischemic episode of more prolonged duration. It has been proposed that during an ischemic episode ADP levels increase and reduce the affinity for ATP which is blocking the K_{ATP} channel, therefore the channel is opened leading to the efflux of K^+ which generates the repolarization of the membrane and the closing of Ca^{2+} channels which prevents the overload of Ca^{2+} in the cardiac cell; yet, many studies on cardioprotection suggest that mitoK_{ATP} may be the final effector of the protection (Grover and Garlid 2000).

The following potassium channels have been described in the inner mitochondrial membrane: ATP-regulated potassium channel (the mitoK_{ATP}); large Ca^{2+}-activated potassium channel (the mitoBK_{Ca} channel), voltage-gated potassium channel (the mitoKv 1.3 channel), and the twin pore TASK-3 potassium channel (Szewczyk et al. 2009). From these, mitoK_{ATP} is the most widely expressed. The subunits that compose the mitochondrial potassium channels are similar to those potassium channels expressed in the plasma membrane and also share the regulation patterns with these channels.

The large conductance Ca^{2+}-regulated potassium (BK_{Ca}) channel is distributed in the plasma membrane of excitable and non-excitable cells. In the first instance, this channel was first identified in smooth muscle, brain and chromaffin cells. The BK_{Ca} channel is a tetramer formed by one α-subunit and the other subunits correspond to β-subunits from different types. Four of these subunits have been cloned $\beta1$-$\beta4$ (Ghatta et al. 2006). The subunits $\beta2$ and $\beta4$ are specifically localized in neurons (Wallner et al. 1999). The expression of the BK_{Ca} channel was identified in the mitochondria from rat brain and human glioma cell line LN229 (Siemen et al. 1999); it is stimulated by Ca^{2+} or the potassium channel opener NS1619 and blocked by charybdotoxin (ChTx) (Kulawiak and Bednarczyk 2005). The identification of the mitoBK_{Ca} in the mitochondria was performed with immunoblots in heart mitochondria against the C-terminal region of the plasma membrane BK_{Ca} channel, identifying a protein of 55 kDa which could be the mitochondrial BK_{Ca} (Sato et al. 2005).

The mitoBK_{Ca} may be cardioprotective; as other mitochondrial channel its opening causes K^+ to flux into the mitochondria along its electrochemical gradient; this produces the depolarization of the membrane and accelerates flavoprotein oxidation, measured as the autofluorescence of mitochondrial flavoprotein (Sato et al. 2005; Hannson et al. 2010) because Ca^{2+} uptake into the mitochondria is a membrane potential dependent process, the reduced influx of Ca^{2+} into the mitochondria prevents its overload avoiding cellular damage and death.

The study of mitoBK_{Ca} in ventricular guinea pig myocytes shows that the activation of the channel by NS1619 is amplified by 8-bromoadenosine-3',5'-cyclic monophosphate (8Br-cAMP). The opening of mitoBK_{Ca} and the influx of K^+ is evaluated by following an increase in flavoprotein oxidation and also with forskolin, a direct activator of adenylate cyclase; however, PKC activating phorbol ester (PMA) did not alter flavoprotein oxidation induced by

NS1619. These results indicate that mitoBK_{ca} channels are modulated by PKA but not by PKC (Sato et al. 2005).

mitoBK_{Ca} is also involved in neuroprotection. With patch-clamp techniques it has been reported that hypoxia inhibits the mitochondrial permeability transition pore by activating the mitoBK_{Ca} channel; this could be interpreted as an anti-apoptotic mechanism (Cheng et al. 2008). It was observed in isolated rat brain mitochondria that the addition of Ca^{2+} to the medium dissipates the mitochondrial membrane potential and increases mitochondrial respiration (Skalska et al. 2009). These effects were blocked with ChTx. Because Ca^{2+} movement depends on mitochondrial membrane potential, K^+ efflux lowers the potential and therefore prevents mitochondrial Ca^{2+} overloading producing a neuroprotective effect (Cheng et al. 2008). Data suggest that the mitoBK_{Ca} could represent a link between cellular-mitochondrial Ca^{2+} signaling and mitochondrial potential reaction to protect the cell (Bednarczyk 2009).

TASK-3 channels are widely distributed and they are found in both excitable and non-excitable cells. These channels are constituted by four transmembrane segments and are active as dimers. TASK channels have an important role in the control of K^+ homeostasis and cell volume; they have been implicated in the modification of the electrical membrane potential, hormonal and neurotransmitter secretion, as well as neuronal and muscular excitability (Lesage and Lazdunski 2000). In melanoma and keratinocyte cells mitochondrial TASK-3 channels were found using immunochemical and molecular biological methods, but its functional properties have not been established (Bednarczyk 2009).

It has been proposed that potassium channels play an important role in cardiac and neuronal protection (O' Rourke 2004; Ardehali 2005), this can be done by preventing the overload of Ca^{2+}; yet, the complete cytoprotective mechanism, and which of the potassium channels described in mitochondria has the major control in the regulation of K^+ mitochondrial homeostasis and volume and therefore the protective effect are data still unknown. The first mitochondrial potassium channel that was described and the most studied is the mitoK_{ATP} channel; it was the first channel identified in the inner membrane of liver and heart mitochondria (Paucek et al. 1992).

Mitochondrial function is essential for tissue protection following ischemia-reperfusion via the production of signaling reactive oxygen species for the activation of protective kinases (Costa et al. 2008). The activation of mitoK_{ATP} is able to maintain the energy exchange with the cytoplasm during stress conditions (Garlid et al. 2006).

During mitochondrial respiration, the heart generates an electrical gradient across the inner membrane of 160-180 mV which makes K^+ diffuse across the mitochondrial inner membrane into the matrix; the efflux and influx of K^+ is kept in a steady state by K^+ efflux via the K^+/H^+ antiporter (Garlid and Paucek 2003). The opening of mitoK_{ATP} is accompanied by osmotic flux of water and the electroneutral transport of anions, mainly phosphate, which leads to a new steady state characterized by three changes: small increase in matrix volume, modest alkalinization and very slight acceleration of respiration (Costa et al. 2006; Kowaltowski et al. 2001). The opening of mitoK_{ATP} occurs during physiological cellular activity through either plasma membrane receptor activation (which signal migrates to mitochondria and phosphorylates the channel) or in response to the opening of K_{ATP} channel as ATP decreases. But when the mitoK_{ATP} remains open, matrix alkalinization is greater producing a partial inhibition of the Complex I and increases the amount of reactive oxygen species (Andrukhiv et al. 2006).

The presence of the mitoK_{ATP} channel in brain mitochondria has been elucidated by biochemical and electrophysiological studies demostrating that mitoK_{ATP} is inhibited by ATP, ADP, and long chain CoA esters, glyburide and 5-hydroxydecanoate (5-HD). The opener of the mitoK_{ATP} channel is diazoxide and the inhibition of ATP is reversed by GTP or GDP, the affinity values ($K_{1/2}$) of the modulators are similar in liver and heart (Bajgar et al. 2001; Garlid et al. 2001). Since the same ligands regulate K_{ATP} channels found in plasma membrane and mitoK_{ATP}, it has been suggested that K_{ATP} and mitoK_{ATP} have a similar protein composition, although the molecular identity of the mitoK_{ATP} channel is still unknown (Bajgar et al. 2001). However, the presence of the mitoK_{ATP} channel has been described in liver, heart, kidney, skeletal muscle, human T-lymphocytes and amoeba mitochondria.

In mitochondria from liver and heart of rat two proteins of 55 and 63 kDa have been purified. These proteins reconstituted in liposomes showed K^+ transport with similar pharmacological response to the plasma membrane K-channel. With these results, it was proposed that mitoK_{ATP} had an inward-rectifying protein Kir 6.1 or Kir 6.2 and the sulfonylurea receptors (SUR) subunits (Inoue et al. 1991). Kir channels play an important role in the maintenance of membrane potential and K^+ homeostasis regulating cellular functions such as membrane excitability, heart rate and vascular tone (Levitan 2009). This family of channels has been divided in seven subfamilies (Kir 1-Kir 7); the classification is based on the sequence homology and their biophysical properties.

Kir 6 has two types of channels, Kir 6.1 and Kir 6.2, which are widely expressed and usually co-assembled with the sulfonylurea receptor (Oliver et al. 2000). The sulfonylurea receptor belongs to the ATP-binding cassette (ABC) transporter and hydrolyzes ATP to transport different substrates out of the cell or into intracellular organelles, but the subunit SUR1 and SUR2 are the only ones of this ABC family transporters that serve as ion channel regulators in the K_{ATP} channel (Aittoniemi et al. 2008).

The proposed Kir 6 subunit as part of the mitoK_{ATP} has not been confirmed. Studies with immunoblotting and immunogold electron microscopy have shown both Kir 6.1 and Kir 6.2 in isolated heart mitochondria (Lacza et al. 2003); Kir 6.1 and Kir 6.2 have also been found in mitochondira of intact rat ventricular myocytes using the marker Mitofluor red (Singh et al. 2003). However, when Kuniyasu et al. (2003) used polyclonal antibodies against the subunits Kir 6.1 or Kir 6.2 an amplified signal was observed in the microsomes but not in mitochondria. Similar results were described by Ng et al. (2010) who also found that Kir 6.1 had a predominant expression in the endoplasmic reticulum with only a modest proportion in mitochondria. Until now, the molecular composition of the K_{ATP} channel found in the mitochondria has not been identified.

Marban et al. (2004) have proposed that the mitoK_{ATP} channel can be conformed by a complex of proteins of which succinate dehydrogenase forms part (Ardehali et al. 2004). This has been suggested because the inhibition of the succinate dehydrogenase enzyme from the Krebs cycle could mimic the effect of cardioprotection observed in the process of ischemic preconditioning, and because some drugs that the activated mitoK_{ATP} channel also inhibited succinate dehydrogenase. This raises the alternative explanation that it is respiratory inhibition rather than channel activity which underlies ischemic preconditioning.

Ardehali et al. (2004) have proposed that succinate dehydrogenase is part of the mitoK_{ATP} channel because this enzyme co-immunoprecipitates with the following proteins from the inner membrane mitochondrial of rat liver mitochondria: mitochondrial ATP-binding cassette protein 1 (mABC1), phosphate carrier (PIC), adenine nucleotide translocator (ANT), and α-subunit ATP synthase (ATPase). These proteins incorporated into proteoliposomes or lipid bilayers, show features and functions similar to those of the mitoK_{ATP} channel. Since there is not a molecular evidence of the composition of the mitoK_{ATP}, the authors suggest that this supercomplex could be the mitochondrial potassium channel.

Many groups have identified different proteins that could be part of the mitoK_{ATP}, but today we do not know its molecular identity. Like other potassium channels, mitoK_{ATP} might have a structure similar to the plasma membrane K_{ATP} with a tissue specific expresion. For example, in pancreas cells the K_{ATP} channel is composed by Kir 6.2/SUR1 (Song and Ashcroft 2001), but in the heart the composition of the plasma membrane channel is Kir 6.2/SUR2A (Babenko et al. 1998). Several authors follow the presence of the expression of the channel based on the mRNA of the Kir 6.1/SUR 2B; for example, KIR 6.1 was detected in rat corpus luteum and placenta but in primates the presence of the K_{ATP} channel has not described (Kunz et al. 2006).

In human beings the role of other ions channel in endocrine cells such as voltage –activated Na^+-channel, Ca^{2+}-activated or the large conductance K^+ channel has been described (Bulling et al. 2000; Kunz et al. 2002). In human ovary a functional K_{ATP} channel was described in plasma membrane; when this channel is opened by stimulation with human chorionic gonadotropin hormone, an increase of progesterone synthesis is observed but inhibition is produced in the presence of glibenclamide (Kunz et al. 2006). Therefore, mitoK_{ATP} appears to be related to steroidogenesis.

The effect of K^+ and stimulation of steroidogenesis by the opening of the K_{ATP} channel is not the only mechanism by which the cell is stimulated by the granulosa cells, it has been reported that the modification of ion concentration stimulates steroidogenesis in different tissues; this could be due to the change of the membrane potential or by a second messengers in the cell (Gore et al. 1984).

Granulosa cells are not the only cells that respond to the stimulation by ions, glomerulosa cells of adrenal glands have as regulators of aldosterone synthesis the hormones angiotensin II (Ang II) and adrenocorticotrophin (ACTH) and K^+. Intracellular increases of K^+ activate two protein kinases, Ca^{2+} -calmodulin-dependent protein kinase (CamK) and diacylglycerol-dependent protein kinase (PKC) (Foster 2004).

The cells respond to ion movement and this could be due to the change in membrane potential or the activation of some kinases. Human placenta exchanges ions across the syncytiotrophoblast. This process is essential for normal fetal homeostasis and humoral and autocrine/paracrine factors are important for the regulation of both tissue homoestasis and nutrient transfer to the fetus (Clarson 2001).

One of these factors is ATP, which in other tissues such as the nervous system and epithelia is important as a paracrine/autocrine regulator of cell function (Gordon 1986; Rathbone et al. 1999); ATP could be released from the cell in response to hypo-osmotic stress or activate the cell via ATP-binding cassette (ABC) transporters (Al-awqati 1995). The human placenta expresses ABC transporters on the microvillous membrane of the syncytiotrophoblast which let ATP out of the cell. In isolated cytotrophoblast cells exposed to extracellular ATP increases intracellular Ca^{2+}; therefore ATP may be an important autocrine/paracrine regulator (Petit and Belisle 1995). Since cytotrophoblast cells respond to the presence of ATP and the K_{ATP} channel functions as a metabolic sensor of the cell, this channel could be expressed in the plasma membrane of cytotrophoblast cells.

Mitochondria from human placenta isolated in the presence of K^+ as iso-osmotic agent showed different morphology with respect to mitochondria isolated with sucrose. Also, some mitochondrial activities were modified in the presence of increasing K^+ concentration such as steroidogenesis and cholesterol incorporation (Martinez et al. 1995).

Cholesterol in steroidogenic tissues is stored in cytoplasmic lipidic droplets, and must be transported to the mitochondria for its transformation into pregnenolone in the mitochondria in response to hormone signals (Christenson and Strauss 2001). In cells from syncytiotrophoblast scarce lipid droplets have been observed, because there is a constant transport of cholesterol from cytoplasm to mitochondria for progesterone synthesis in a non-acute hormone dependent system.

Cholesterol is one of the main lipidic components of plasma membrane in all mammalian cells and it is essential for cell function and growth, an excess of cholesterol is cytotoxic and is associated with the development of cardiovascular disease, but changes in membrane cholesterol affect the activity of some membrane receptors and ion channels such as K^+ (Levitan 2009). An increase of cholesterol in plasma membrane result in a significant decrease in the conductance currents of Kir 2, but not in the open probability of the channel; the change in cholesterol concentration decreases the number of active channels present in the cell. The effect of the cholesterol on the channel activity of Kir 6 is still in controversial; some studies show that the activity of Kir 6 is enhanced by plasma hypercholesterolemia (Mathew et al. 2001) and other studies show that hypercholesterolemia blocks the activity of the Kir 6 channel (Genda et al. 2002).

In human placenta mitochondria progesterone synthesis increases in the presence of isocitrate (García-Pérez et al. 2002); when K^+ is added, the steroidogenesis increases 5-6 times (Milan et al. 2010); similar results are observed when glibenclamide or other mitoK_{ATP} inhibitor is added to the medium. The channel used to transport K^+ in human placenta mitochondria has not been described, but the use of inhibitors such as glibenclamide, TPP or quinine which block K^+ transport, suggests that mitoK_{ATP} is involved. The identification of the channel associated with K^+ transport in placenta mitochondria was performed with antibodies against the Kir 6.1 subunit by Western Blot. The confirmation of its presence in human placenta mitochondria was supported by the comparison of the results obtained in isolated rat mitochondria in which Kir 6.1 was identified by Garlid and Pauchek (2001) (Milan et al. 2010).

The expression of potassium mitochondrial channels could be tissue specific and more than one of these channels could be expressed in mitochondria to ensure K^+ homeostasis, mitochondria volume, and the membrane potential necessary for the mitochondrial activity and, therefore, cell survival. More studies should be performed to clarify which potassium channels are expressed in each tissue and the mechanism that controls their activity in the mitochondria and the cell.

Acknowledgments

This work was partially supported by Grants IN203006 and IN217609 from Dirección General de Apoyo al Personal Académico de la Universidad Nacional Autónoma de México. The authors thank Dr. José Luis Pérez-García for reviewing the correct use of English in this manuscript.

References

Aittoniemi J, Fotinou J, Craig TM, de Wet H, Proks P, Ashcroft FM. SUR1: a unique ATP-binding cassette protein that functions as an ion channel regulator. *Philos Trans R Soc Lond B Biol Sci* (2008) 364:257-267.

Al-awqati Q. Regulation of ion channels by ABC transporters that secrete ATP. *Science* (1995) 269:805-806.

Alekseev AE, Reyes S, Yamada S, Hodgson-Zingman DM, Sattiraju S, Zhu Z, Sierra A, Gerbin M, Coetzee WA, Goldhamer DJ, Terzic A, Zingman LV. Sarcolemmal ATP-sensitive control energy expenditure determining body weight. *Cell Metab.* (2010) 11:58-69.

Andrukhiv A, Costa AD, West IC, Garlid KD. Opening mitoKATP increases superoxide generation from complex I of the electron transport chain. Am *J. Physiol.* (2006) 291:H2067-2074.

Ardehali H. Cytoprotective channels in mitochondria. *J. Bioenerg. Biomembr.* (2005) 37:171-177.

Ardehali H, Chen Z, Ko Y, Mejia-Alvarez R, Marban E. Multiprotein complex containing succinate dehydrogenase confers mitochondrial ATP-sensitive K^+ channel activity. *Proc. Natl. Acad. Sci. USA* (2004) 101:11880-11885.

Babenko AP, Gonzalez G, Aguilar-Bryan L, Bryan J. Reconstituted human cardiac K_{ATP} channels: Functional identity with the native channels from the sarcolemma of human ventricular cells. *Circ. Res.* (1998)83:1132-1143.

Bajgar R, Seetharaman S, Kowaltowski AJ, Garlid KD, Paucek P. Identification and properties of a novel intracellular (mitochondrial) ATP-sensitive potassium channel in brain. *J. Biol. Chem.* (2001) 26: 33369-33374.

Bednarczyk P. Potassium channels in brain mitochondria. *Acta Biochem. Pol.* (2009) 56:385-392.

Brayden JE. Functional roles of K_{ATP} channels in vascular smooth muscle. *Clin. Exp. Pharmacol. Physiol.* (2002) 29:312–316.

Bulling A, Berg FD, Berg U, Duffy DM, Stouffer RL, Dissen GA, Ojeda SR, Mayerhofer A. Identification of an ovary voltage-activated Na^+-channel type: Hints to involvement in luteolysis. *Mol. Endocrinol.* (2000) 14:1064-1074.

Cheng Y, Gu XQ, Bednarczyk P, Wiedemann FR, Haddad GC, Siemen D. Hypoxia increases activity of the BK-channel in the inner mitocondrial membrane and reduces activity of the permeability transition pore. *Cell Physiol. Biochem.* (2008) 1656:46-56.

Clarson LH, Greenwood SL, Myolona P, Sibley CP. Inward rectifying K^+ current amd differentiation of human placental cytotrophoblast cells in culture. *Placenta* (2001) 22:328-336.

Costa AD, Jakob R, Costa CL, Andrukhiv K, West IC, Garlid KD. The mechanism by which the mitochondrial ATP-sensitive K^+ channel opening and H_2O_2 inhibit the mitochondrial permeability transition. *J. Biol. Chem.* (2006) 281:20801-20808.

Costa AD, Pierre SV, Cohen MV, Downey JMZ, Garlid KD. cGMP signalling in pre- and post-conditioning: the role of mitochondria. *Cardiovac. Res.* (2008) 344-352.

Christenson LK and Straus JF[3rd]. Steroidogenic acute regulatory protein: an update on its regulation and mechanism of action. *Arch. Med. Res.* (2001) 32:576-586.

Doyle DA, Cabral JM, Pfuetzner RA, Kuo AL, Gulbs JM, Chait BT and Mackinnon R. The structure of the potassium channel: molecular basis of K^+ conduction and selectivity. *Science* (1998) 280:69-77.

Foster RH. Reciprocal influences between the signaling pathways regulating proliferation and steroidogenesis in adrenal glomerulosa cells. *J. Mol. Endo.* (2004) 32:893-902.

García-Pérez C, Pardo JP, Martínez F. Ca^{2+} modulates respiratory and steroidogenic activities of human term placental mitochondria. *Arch. Biochem. Biophys* (2002) 405:104-111.

Garlid KD and Paucek P. The mitochondrial potassium cycle. *IUBMB Life* (2001) 52:153-158.

Garlid KD and Paucek P. Mitochondrial potassium transport: the K^+ cycle. *Biochim. Biophys. Acta* (2003)1606:23-4.

Garlid KD, Puddu PE, Pasdois P, Costa AD, Beauvoit B, Criniti A, Tariosse L, Diolez P, Dos Santos P. Inhibition of cardiac contractility by 5-hydroxydecanoate and tetraphenylphosphonium ion: a possible role of mitoKATP in response to inotropic stress. *Am. J. Physiol. Heart Circ. Physiol.* (2006) 291: H152–H160.

Genda S, Miura T, Miki T, Ichikawa Y, Shimamoto K. KATP cannel openning is an endogenous mechanism of protection against the no-reflow phenomenon but its function is compromised by hypercholesterolemia. *J. Am. Col. Cardiol.* (2002) 40:1339-46.

Ghatta S, Nimmagadda D, Xu X, O'Rourke ST. Large-conductance, calcium-activated potassium channel: Structural and functional implications. *Pharmacol. Ther.* (2006) 110:103-116.

Gordon JL. Extracellular ATP: Effects, source and fate. *Biochem. J.* (1986) 233:309-319.

Gore SD and Behrman HR. Alteration of transmembrane sodium and potassium gradients inhibit the action of luteinizing hormone in the lutheal cell. *Endocrinology* (1984) 114:2020-2031.

Grover GJ and Garlid KD. ATP-Sensitive potassium channels: A review of their cardioportective phamarcology. *J. Mol. Cell Cardiol* (2000) 32:677-695.

Hannson MJ. Morota S, Teilum M, Mattiasson G, Uchino H, Elmér E. Increased potassium conductance of brain mitochondria induces resistence to permeability transition by enhancing matrix volume. *J. Biol. Chem.* (2010) 285:741-750.

Inoue I, Nagase H, Kishi K, Higuti T. ATP-sensitive K^+ channel in the mitochondrial inner membrane. *Nature* (1991) 352:244-247.

Jenkinson DH. Potassium channels – multiplicity and challenges. *Br. J. Pharmacol.* (2006) 147:S63-S71.

Kowaltowski AJ, Seetharaman S, Paucek P, Garlid KD. Bioenergetic consequences of opening the ATP-sensitive K^+ channel of heart mitochondria. *Am. Physiol. Heart Circ. Physiol.* (2001) 280:H649-H657.

Kulawiak B and Bednarczyk P. Reconstitution of brain mitochondria inner membrane into planar lipid bilayer. *Acta Neurobiol. Exp.* (2005) 65:271-276.

Kuniyasu A, Kaneko K, Kawahara K, Nakayama H. Molecular assembly and subcellular distribution of ATP-sensitive potassium cannel proteins in rat hearts. *Febs Lett* (2003) 552:259-263.

Kunz L, Thalhammer A, Berg FD, Berg U, Duffy DM, Stouffer RL, Dissen GA, Ojeda SR, Mayerhofer A. Ca^{2+} -activated large conductance K^+ channel in the ovary: Identification, characterization, and functional involvement in steroidogenesis. *J. Clin. Endocrinol. Metab.* (2002) 87:5566-5574.

Kunz L, Richter JS, Mayerhofer A. The adenosine 5'-triphosphate-sensitive potassium channel in endocrine cells of the human ovary: Role in membrane potential generation and steroidogenesis. *J. Clin. Endocrinol. Metab.* (2006) 91:1950-1955.

Lacza Z, Snipesa JA, Kisa B, Szabó C, Groverd G, Busijaa DW. Investigation of the subunit composition and the pharmacology of the mitochondrial ATP-dependent K^+ channel in the brain. *Brain Res.* (2003) 994:27– 36.

Lesage F and Lazdunski M. Molecular and functional properties of two-pore-domain potassium channels. *Am. J. Physiol. Renal. Physiol.* (2000) 279:F793-F801.

Levitan I. Cholesterol and Kir channels. *IUBMB Life* (2009) 61:781-790.

Martinez F, Pardo JP, Flores-Herrera O, Espinosa-García MT.The effect of osmolarity on human placental mitochondria function. *Int. J. Biochem. Cell Biol.* (1995) 27:795-803.

Mathew V and Lerman A. Altered effects of potassium channel modulation in the coronary circulation in experimental hypercholesterolemia. *Atherosclerosis* (2001) 154:329-335.

Milan R, Flores-Herrera O, Espinosa-Garcia MT, Olvera-Sanchez S, Martinez F. Contribution of potassium in human placental steroidogenesis. *Placenta* 31:860-866.

Ng KE, Schwarzer S, Duchen MR, Tinker A. The Intracellular Localization and Function of the ATP-Sensitive K^+ Channel Subunit Kir 6.1. *J. Membr. Biol.* (2010) 234:137–147.

O'Rourke B. Evidence for mitochondrial K^+ channels and their role in cardioprotection. *Circ. Res.* (2004) 94: 420–432.

Oliver D, Baukrowitz T, Fakler B. Polyamines as gating molecules of inward-rectifier K^+ channels. *Eur. J. Biochem.* (2000) 267: 5824-5829.

Olson TM and Terzic A. Human K_{ATP} channelopathies: Diseases of metabolic homeostasis. *Eur. J. Physiol.* (2010) 460:295-306.

Paucek P, Mironova G, Mahdi F, Beavis AD, Woldegiorgis G, Garlid KD. Reconstitution and partial purification of the glibenclamide-sensitive, ATP-dependent K^+ channel from rat liver and beef heart mitochondria. *J. Biol. Chem.* (1992) 267:26062-26069.

Petit A and Belisle S. Stimulation of intracellular calcium concentration by adenosine triphosphate and uridine 5'-triphosphate in human term placental cells: Evidence for purinergic receptors. *J. Clin. Endocrinol. Metab.* (1995) 80:1809-1815.

Rathbone MP., Middlemise PJ., Gysber JW., Andrew C., Herman MAR., Reed JK., Ciccarelli R., Iorio PI., Caciagli F. Trophic effects of purines in neurons and glial cells. *Prog. Neuribiol.* (1999) 59:663-690.

Sato T, Saito T, Saegusa N, Nakaya H. Mitochondrial Ca^{2+}-activated K^+ channels in cardiac myocytes: a mechanism of the cardioprotective effect and modulation by protein kinase A. *Circulation* (2005) 111:198-203.

Siemen D, Loupatatzis C, Borecky J, Gulbins E, Lang F. Ca^{2+}-activated K channel of the BK-type in the inner mitochondrial membrane of a human glioma cell line. *Biochem. Biophys. Res. Commun.* (1999) 257: 549–554.

Singh H, Hudman D, Lawrence CL, Rainbow RD, Lodwick D, Norman RI. Distribution of Kir6.0 and SUR2 ATP-sensitive potassium channel subunits in isolated ventricular myocytes. *J. Mol. Cell Cardiol.* (2003) 35:445-449.

Skalska J, Bednarczyk P, Piwońska M, Kulawiak B, Wilczynski G, Dołowy K, Kudin AP, Kunz WS, Szewczyk A. Calcium ions regulate K uptake into brain mitochondria: The evidence for a novel potassium channel. *Int. J. Mol. Sci.* (2009) 10:1104-1120.

Song DK and Ashcroft FM. ATP modulation of ATP-sensitive potassium channel ATP sensitivity varies with the type of SUR subunit. *J. Biol. Chem*. (2001) 76:7143-9.

Szewczyk A, Jarnuszkiewicz W, Kunz WS. Mitochondrial potassium channels. *IUBMB Life* (2009) 61:134-43.

Wallner M, Meera P, Toro L. Molecular basis of fast inactivation in voltage and Ca^{2+}-activated K^+ channels: A transmembrane beta-subunit homolog. *Proc. Natl. Acad. Sci. USA* (1999) 96: 4137–4142.

Zhou HX. The gates of ion channels and enzymes. *Trends Biochem. Sci.* (2010) 35:179-185.

In: Potassium Channels
Editor: Danielle S. Fonseca

ISBN: 978-1-61324-880-5

Chapter IV

Slow Potassium Channel Dysfunction in Amyotrophic Lateral Sclerosis and Its *In Vivo* Evaluation by Threshold Tracking

Hiroyuki Nodera*
Department of Neurology,
Beth Israel Deaconess Medical Center, Boston, MA, U. S.

Abstract

Amyotrophic lateral sclerosis (ALS) is a neurodegenerative disease that predominantly affects upper and lower motor neuron systems. In addition to muscle wasting and weakness, fasciculation is a characteristic clinical and electrophysiological feature in ALS that suggests abnormal excitability of motor nerves. Nerve hyperexcitability itself is neurotoxic, because the reverse action of calcium-sodium exchanger increases the intracellular calcium concentration, that leads to cell death. Recently, ion channel dysfunction has been reported in sporadic and familial ALS patients and model animals, such as smaller slow potassium current and greater persistent sodium current, both of which contribute to the

* Correspondence to: Hiroyuki Nodera, MD Department of Neurology, Beth Israel Deaconess Medical Center, Shapiro 8, 330 Brookline Ave, Boston, MA 02215, e-mail: hnodera@bidmc.harvard.edu, phone: 617-667-4382. fax: 617-667-3175

increased nerve excitability. Threshold tracking is a non-invasive in vivo neurophysiologic test that can analyze such ion channel functions. In this chapter, the method of threshold tracking and its findings are discussed. A new class of antiepileptic agent, a slow potassium channel opener, such as retigabine, demonstrated a hyperpolarizing stabilizing effect of the membrane potential. Thus, the slow potassium channel may become a promising target for neurodegenerative diseases by its neuroprotective effect.

Abnormal Nerve Excitability in Amyotrophic Lateral Sclerosis (ALS)

Amyotrophic lateral sclerosis (ALS) is a neurodegenerative disease which mainly affects the upper and lower motor neuron systems. Some of the patients have known causes, such as mutation of superoxide dismutase 1 (SOD1) and optineurin (Maruyama et al. , 2010, Turner and Talbot, 2008). However, in the majority of the patients with ALS, the cause of the disease is largely unknown. Clinically, ALS is characterized by upper motor neuron signs such as increased tendon reflexes and pathological reflexes such as Babinski sign as well as lower motor neuron signs such as muscle wasting and fasciculation.

Of those, fasciculation, which is defined as involuntary contraction or twitching of groups of muscle fibers and is considered to be a very unique feature of ALS, although other conditions can demonstrate it. Fasciculation is considered to be associated with abnormally increased excitability of the motor system.

The origin of the fasciculation in ALS has been extensively studied and at least two different sites of origin have been reported. Kleine et al.recently reported that high-density surface electromyography (EMG) can differentiate two patterns of fasciculation potentials (FPs) from neuronal and axonal origins (Kleine et al. , 2008). de Carvalho, et al. suggested that FP arise proximally in the limbs in the early disease course and distally later. Early in the disease, FP is stable in its rhythm and morphology, but late in the disease FP becomes unstable (de Carvalho and Swash, 1998). It is not uncommon to observe FP in a strong limb in ALS without noticeable weakness and muscle atrophy, thus it is possible that abnormal excitability precedes axonal and neuronal loss in the early stage of ALS.

Ion Channel Dysfunction in ALS

Since FP implies spontaneous firing of a motor neuron or a motor axon, an underlying cause for the increased nerve excitability has been considered. Nerve excitability is defined by multiple factors; a static component determining the resting membrane potential and a dynamic component determining the behavior of the nerve during the depolarization and repolarization after firing of the nerve.

A resting membrane potential is calculated with the Goldman-Hodgkin-Katz equation which includes the permeability of ions, the intra- and extra-cellular concentrations of ions. Practically, intra- and extra-cellular concentrations of Na^+ and K^+ largely determine the resting membrane potential. Unless significant dehydration or renal insufficiency exists, such electrolyte abnormality is unlikely in ALS, thus abnormal nerve excitability due to the abnormal static component is unlikely.

Rather, abnormal functions of various Na^+ and K^+ channels have been reported to be associated with abnormal nerve excitability in ALS. Here, brief introduction of the axonal Na^+ and K^+ channels and their dysfunction in ALS are discussed.

1) *Voltage-gated Na^+ channels* (VGSC) are clustered in the nodes of Ranvier which open to depolarize the membrane to allow saltatory transmission in a myelinated axon (Figure 1). Abnormality of VGSC has been reported by a few groups in ALS. Gunasek, et al. exposed the cerebrospinal fluid (CSF) of sporadic ALS patients to the rat spinal cord cultures and the spinal cord sections. They observed a decrease in the expression of $Na_V1.6$ and $K_V1.6$ channels in motor neurons in ALS-CSF treated group (Gunasekaran et al. , 2009). Zona, et al. studied the functionality of VGSC in dissociated motor neurons in culture from a mouse model of familial ALS which has mutated Cu-Zn superoxide dismutase (G93A) and observed that the recovery from fast inactivation was significantly faster in G93A motor neurons than that in control (Zona et al. , 2006).
2) *Persistent Na^+ channel.* In addition to the fast inactivating Na^+ current produced by voltage-gated Na^+ channels which are associated with generation of action potential, the Na^+ channel can give rise to a non-inactivating or persistent Na^+ current. It was established that the persistent Na^+ current has functional significance in setting the membrane potential in a subthreshold range.

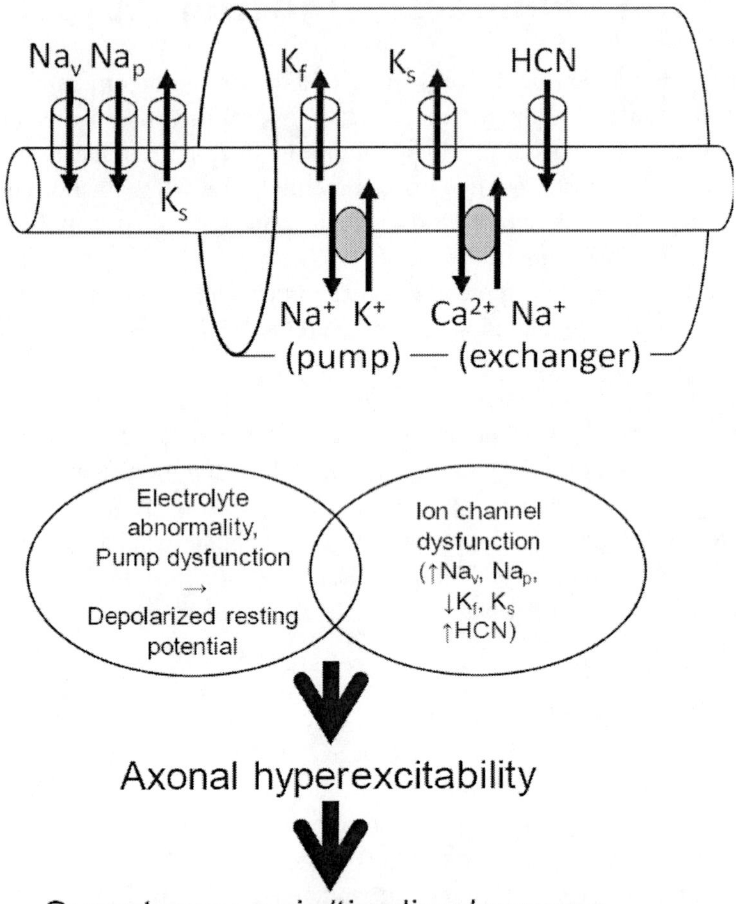

Figure 1. Localization of ion channels and ion pumps in the peripheral myelinated nerve (top). Fast axonal K^+ channels (K_f) are primarily located in internodes, whereas few K_f are located in the nodes of Ranvier. Because of the low concentration, K_f can be ignored when modeling action potential. Slow axonal K^+ channels (K_s), on the other hand, are more concentrated at the node than in the internode. The kinetics of K_s is too slow to contribute to the generation of action potential. K_s, however, help determine the resting membrane potential and contribute to accommodate to long-lasting depolarizing stimuli. The bottom figure shows the contributing factors for axonal hyperexcitability. Abnormal resting membrane potential due to factors such as electrolyte abnormality and pump dysfunction and ion channel dysfunction provoke axonal hyperexcitability resulting in symptoms such as tingling, pain, and cramps. Because ion channel dysfunction also contributes to resting membrane potential, these factors overlap significantly.

Therefore, the membrane potential becomes more depolarized with greater persistent Na^+ current and vice versa (Kiss, 2008). Tamura, et al. utilized the in vivo electrophysiological analysis called latent addition and reported that persistent Na^+ current was increased in sporadic ALS (Tamura et al. , 2006). Similarly increased persistent Na+ conductances were reported in SOD-1 mutated familial ALS (Vucic and Kiernan, 2010). Pieri, et al. reported that the cultured cortical neurons from G93A ALS model mouse showed the significantly higher persistent Na+ current (Pieri et al. , 2009). Taken together, the abnormally increased persistent Na+ current in sporadic and familial ALS appears to contribute to the hyperexcitability of motor axons.

3) *Fast K^+ channel.* The voltage-gated K^+ channels are located in the juxtaparanode region which is mechanically separated from a node of Ranvier by paranodes. Because of the localization, the voltage-gated K^+ channels do not account significantly for the production of saltatory transmission. In paranodal demyelination as seen in inflammatory demyelinating neuropathy, the voltage-gated K^+ channels become exposed and contribute to hyperpolarizing the membrane potential after membrane depolarization. Given the already abnormal safety factor of transmission, this hyperpolarizing effect by the voltage-gated K^+ channels limits successful saltatory transmission. A FDA-approved fast K^+ channel blocker, 4-Aminopyridine (dalfampridine), has shown clinical effect of improving the gait function in multiple sclerosis (Goodman et al. , 2010).
4) *Slow K^+ channel.* Physiological experiments have identified a characteristic time- and voltage-dependence that results in the 'clamping' of the membrane potential if the channel is exposed to an excitatory stimulus. This was called the M current (Delmas and Brown, 2005). The current is characterized by slow kinetics of activation and deactivation, no inactivation, and blockage by TEA but not by 4-aminopyridine. The M channel functions as a 'brake' on repetitive action potential discharges and, as such, has a key role in regulating the excitability of various central and peripheral neurons. The responsible channel had not been identified for many years until 1998 when Wang, et al. identified the KCNQ2 and KCNQ3 potassium channel subunits contribute to the native M-current (Wang et al. , 1998). Based on the new nomenclature of potassium channels (Gutman et al. , 2003), the subunits are now classified as Kv7

channels. Because mutations in the gene encoding the Kv7 channels cause neonatal epilepsy and myokymia, Kv7 has been considered to regulates the excitability of central and peripheral nerves. Devaux, et al. identified the localization of Kv7.2 (KCNQ2) and Kv7.3 (KCNQ3) at nodes of Ranvier and the initial segments (Devaux et al. , 2004). Schwarz, et al. further (Schwarz et al. , 2006) further demonstrated the following: (1) immunostaining showed that the Kv7.2 was located diffusely at the nodes regardless of the fiber size, whereas the Kv7.3 was predominantly present in the thin axons, (2) the opening of nodal Kv7 channel contributes to hyperpolarizing shift of the membrane potential, and (3) blocking of Kv7 channels by its antagonist XE991eliminated all nerve excitability functions previously attributed to the slow potassium current (refer to nerve excitability testing in the following section) and the animal showed increased excitability manifesting as fasciculations (Schwarz, Glassmeier, 2006). A few compounds have been identified as agonists for Kv7 channels, or called as the slow potassium channel openers. Retigabine is the classic example which has shown clinical efficacy for partial epilepsy due to its stabilizing effects (Brodie et al. , 2010). Flupirtine was another Kv7 channel opener and has been marketed in Europe since 1984 for treatment of pain, but has not been introduced to the U.S. Flupirtine was reported to have effects on peripheral nerve excitability similar to retigabine (Sittl et al. , 2010). Mutation of slow K^+ channel causes various neurological diseases, but commonly characterized by hyperexcitability (Maljevic et al. , 2008, Maljevic et al. , 2010). Neonatal epilepsy caused by Kv7.2 and Kv7.3 channel mutations is a rare, autosomal dominant epilepsy syndrome with a penetrance of >80%, starting within the first days of life with frequent partial and secondary generalized seizures, often in clusters, that disappear spontaneously after several weeks to months. The risk of recurring seizures later in life is about 15%. In most cases, the psychomotor development is described to be normal; however, several cases of patients with mental retardation and difficult-to-treat epilepsies have also been described (Maljevic, Wuttke, 2010). Peripheral nerve hyperexcitability (PNH) caused by voltage sensor Kv7.2 mutations clinically manifest as spontaneous and continuous muscle movements, such as myokymia (undulating movements of distal skeletal muscle), fasciculations, cramps, or other symptoms. While autoantibodies directed against several voltage-gated potassium channels (Kv1.1,

KV1.2, and KV1.6) represent the most common pathophysiology of PNH, mutations in two K+ channel genes, KCNA1 (encoding KV1.1) and KCNQ2 (encoding Kv7.2), have been demonstrated to cause PNH with episodic ataxia type 1 (KCNA1) (Hart et al. , 2002).

In Vivo Nerve Excitability Testing

An electrophysiologic test is one of the most commonly utilized tools for research and clinical study of peripheral nerve diseases. Testing on axonal ion channel functions was only possible by laboratory techniques such as single-cell recording by patch clamp and *in vitro* nerve preparation with single axon recording. Nerve conduction study (NCS) is particularly commonly utilized in a clinical setting because of its non-invasiveness and high reproducibility. In NCS, diagnostic parameters include a conduction velocity and response amplitude, representing integrity of myelins and axons, respectively. Although NCS is powerful in detecting peripheral nerve damage, NCS is not particularly useful in assessing the excitability of the nerve fiber and ion channel functions. Another neurophysiological method, nerve excitability testing (NET), has history even longer than nerve conduction study. A classic form of NET was to apply various strengths of electric stimuli over a nerve to detect its stimulating threshold. However, such a simple method provides only limited information on excitability. Subsequent parameters were developed such as chronaxie and rheobase that infer gross excitability of the axonal membrane. The utilization of NET has skyrocketed since 1970, when Bergmans described increased stimulation threshold by applying impulse trains that induces transient hyperpolarization, which later found due to activation of slow K+ channels. Since then, various stimulating techniques have been introduced to elucidate axonal functions, but the approach was limited to the normal animals and human subjects because of difficulty of setting up the recording system and long recording time. An automated recording program developed by Prof. Hugh Bostock in London 10 years ago has allowed in vivo recording of multiple nerve excitability measures within 15 minutes (Kiernan et al. , 2000). This convenient *in vivo* NET applicable to animals and humans quickly expanded the knowledge of nerve excitability in various peripheral nerve diseases including ALS (Nodera and Kaji, 2006).

Nerve excitability changes following conduction of a single nerve impulse before returning to its resting state called recovery cycle (RC) (Figure 2).

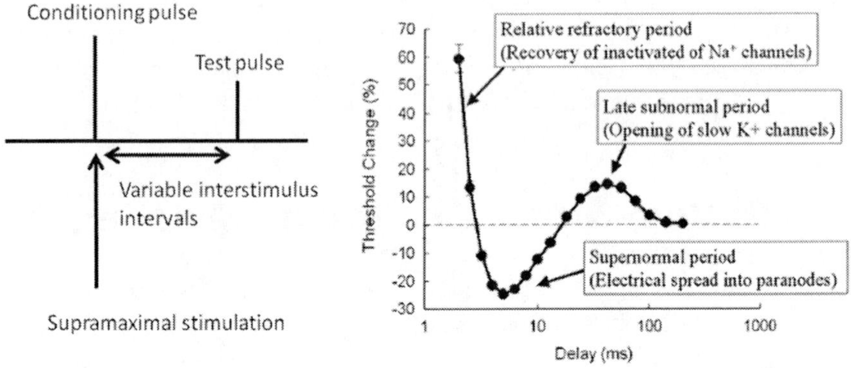

Figure 2. The recording principle of recovery cycle (RC). Double conditioning-test electrical stimulation with various intervals detects changes in the threshold current needed to produce a predetermined motor or sensory respond amplitude (compound muscle action potential (CMAP) or compound nerve action potential (CNAP)) after the conditioning stimulus. By this principle, this method cannot record the absolute refractory period, when the second CMAP or CNAP cannot be elicited regardless of the strength. The relative refractory period, usually observed for 3-4 ms, occurs as the inactivated Na^+ channels recover. It is followed by supernormal period, caused by paranodal spreading of the electric charges. Subsequently, the late subnormal period is present when activation of slow K+ channels hyperpolarize the membrane potential and lower the nerve excitability.

In a normal human peripheral nerve, RC has three phases; refractory period, superexcitability, and late subexcitability. Refractory period is further divided into two; absolute and relative refractory period (ARP and RRP). ARP is the brief period immediately after the test stimuli when the second firing is unable to be achieved regardless of the strength of thepulse. This is because of the inactive state of voltage-gated Na^+ channel. After ARP, RRP follows when strong electric stimulation can provoke the second firing. These two parameters indicate the opening and inactivation of voltage-gated Na+ channels. Follwing the refractory period, superexcitability appears when weaker pulse can prove the same response level as the baseline response. This suggests transient axonal hyperexcitability, maximum approximately at 7 ms after the firing. Superexcitability occurs because focal membrane depolarization at the node of Ranvier slowly spreads into the adjacent paranodes and internodes to "warm" the nodal region to help the second firing to occur easily. Because fast K^+ channels predominantly locate in the internode, the hyperactive fast K^+ channels "cools down" the axonal membrane after the first firing and the threshold for the second firing becomes greater. On the contrary, if the fast K+ channels is dysfunctional, suppressive force of membrane depolarization around the nodal region is decreased, such that superexcitability becomes more prominent and repetitive firing may occur. In neuromyotonia (Isaacs syndrome), there is antibody against voltage-gated K+ channels and its neurophysiologic hallmark if neuromyotonia, which is a very high frequency repetitive discharges from the axonal origin (Arimura et al. , 2006). Late subexcitability appears approximately between 20 and 100 ms following the single nerve impulse and maximizes its value at about 40-50 ms following the impulse (Bostock et al. , 1998, Nodera and Kaji, 2006). It reflects transient membrane hyperpolarization due to slow potassium current (IKs) at the nodes of Ranvier by opening of the slow potassium channels called KCNQ2 and KCNQ3. Late subexcitability can be amplified with a train of multiple conditioning stimuli, as reported by Bergmans called H1 (first hyperpolarizing afterpotential) (Bergmans, 1970). Schwarz, et al. applied a train of seven conditioning stimuli to rats and reported the usefulness of H1 to identify the in vivo electrophysiologic effects of XE991, a slow potassium channel inhibitor (Schwarz, Glassmeier, 2006).

Significance of Abnormal Slow Potassium Current in ALS

Recently, two groups independently reported the impaired function of IKs in amyotrophic lateral sclerosis. Kanai, et al. reported the nerve excitability changes along with the clinical progression (Kanai et al. , 2006). First, persistent Na+ conductance increases, possibly associated with collateral sprouting, and then K+ conductances decline, both of which cause axonal hyperexcitability. Vucic, et al. reported the similar results (Vucic and Kiernan, 2006). Our ALS patients showed the similar abnormalities in nerve excitability. The pathophysiology of impairment of IKs in ALS has yet to be clarified, but should include some of the hypotheses. First, ALS is known to impair axonal transport (De Vos et al. , 2008). Transport of either KCNQ 2 or KCNQ3 channels themselves or their trophic factors may be decreased. Second, various abnormalities of gene expression have been reported, including RNA editing. Either mechanism may decrease the numbers of expressed KCNQ channels on the axonal surface, resulting in decreased IKs.

Figure 3 demonstrates the abnormal slow K+ current in ALS in our patients (Shibuta et al. , 2010).

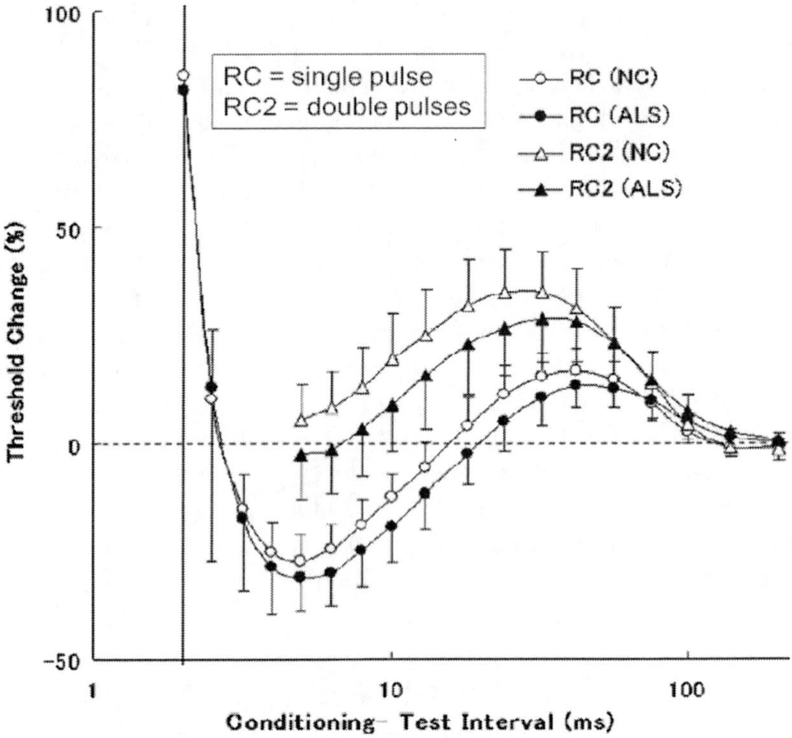

Figure 3. Recovery cycle with single and double conditioning pulses (RC and RC2, respectively) in the ALS and normal control groups (mean ± SD). Within the group, greater threshold changes in late subexcitability by RC2 than by RC were obtained. Of note, the ALS group showed lower threshold changes in late subexcitability than the normal group by both RC and RC2, but the difference between ALS and normal control groups were more noticeable by RC2 (Shibuta, Nodera, 2010) (modified figure).

We performed RC in 22 patients with ALS who met probable or definite revised El Escorial criteria and 22 closely age-matched normal control subjects. The ALS group comprised of 12 men and 10 women (Mean age ± SD: 62.6 ± 8.6 years, range 43-80) and the control group comprised of 12 men and 10 women (Mean age ± SD: 59.9 ± 8.9 years, range 52-78). The averaged maximum values of late subexcitability was smaller in ALS (13.3 ± 4.4%) than in controls (16.0 ± 4.8%) (P = 0.046). By applying double conditioning pulses (RC2), the subexcitability almost doubled in ALS (29.4 ± 9.5%) and controls ((34.9 ± 9.5%) with smaller P value than RC (P = 0.018). Figure 4 shows data from an ALS patient whose late subexcitability was severely decreased by RC, which provided difficulty identifying a single peak.

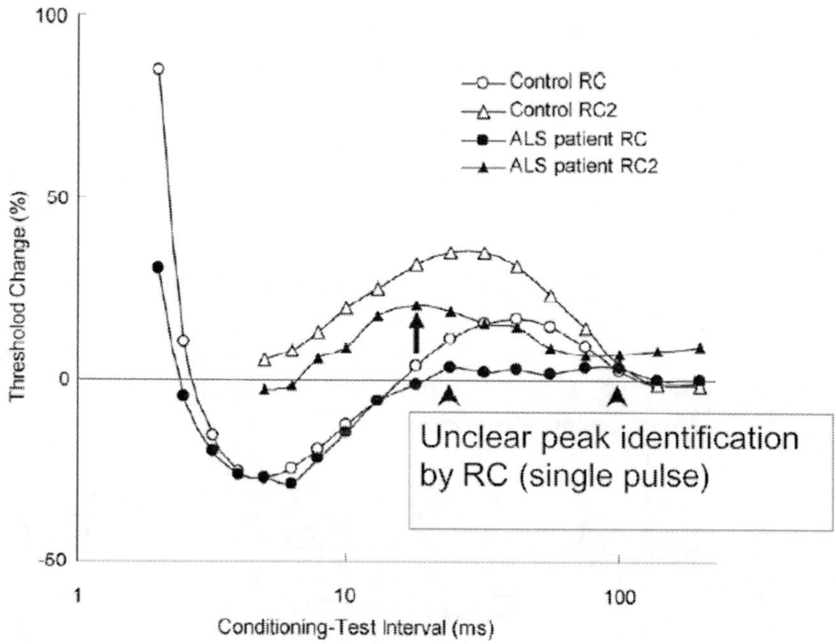

Figure 4. Recovery cycle with single and double conditioning pulses (RC and RC2, respectively) in one of the ALS patients (filled circles and filled triangles) in comparison to control curves (open circles and open triangles). The significantly decreased and flat late subexcitability in RC produces difficulty identifying its peak threshold change. In this example, identical maximum threshold changes were obtained at two different conditioning-test intervals by RC (arrowheads), whereas RC2 of the same patient shows a clear peak (arrow)(Shibuta, Nodera, 2010)(modified figure).

The arrowheads show the two identical peaks. On contrary, the double stimulation (RC2) provided a single peak (arrow), resulting in more accurate measures of slow K+ current. The RC2 was well tolerated by all the subjects. In animals under anesthesia, up to 7 trains of stimuli is possible to further enhance the slow K+ current, however, this is too painful in awake individuals to tolerate in clinical setting.

Therapeutic Implication

The etiopathophysiology of impaired slow K^+ current in ALS has remained elusive. Little is known whether this abnormality develops in the presymptomatic stage or this abnormality is a secondary feature induced by axonal damage. However, given the fact that other neurological diseases affecting motor axon loss such as diabetic neuropathy, Guillain-Barré syndrome, and other lower motor neuron diseases have not be reported to share the feature, impairment of slow K^+ function is probably unrelated to simple axonal degeneration, but rather an unique feature in ALS. Therefore, slow K+ channelopathy appears to have etiological implication. On the other hand, even without clarifying the etiopathogenesis of impaired slow K^+ current in ALS, pharmacological intervention to reverse the abnormality may be clinically meaningful. This is because fasciculations, a characteristic phenomenon in ALS shows the underlying abnormal nerve excitability. Excessive nerve excitability causes unnecessary stress to mitochondria that may facilitates neurodegeneration, as reported in sodium channelopathy (Waxman, 2006).

Conclusion

The abnormally decreased slow K^+ current is present in ALS. This is detectable by non-invasive *in vivo* electrophysiology called threshold tracking. It is to be elucidated whether pharmacological intervention by slow K^+ opener and other agents to reverse the dysfunction has neuroprotective effects.

References

Arimura K, Ng AR, Watanabe O. Immune-mediated potassium channelopathies. *Suppl. Clin. Neurophysiol.* 2006;59:275-82.

Bergmans J. *The Physiology of Single Human Nerve Fibres*. Vander, Belgium: University of Louvain. 1970.

Bostock H, Cikurel K, Burke D. Threshold tracking techniques in the study of human peripheral nerve. *Muscle Nerve*. 1998;21:137-58.

Brodie MJ, Lerche H, Gil-Nagel A, Elger C, Hall S, Shin P, et al. Efficacy and safety of adjunctive ezogabine (retigabine) in refractory partial epilepsy. *Neurology*. 2010;75:1817-24.

de Carvalho M, Swash M. Fasciculation potentials: a study of amyotrophic lateral sclerosis and other neurogenic disorders. *Muscle Nerve*. 1998;21:336-44.

De Vos KJ, Grierson AJ, Ackerley S, Miller CC. Role of axonal transport in neurodegenerative diseases. *Annu. Rev. Neurosci.* 2008;31:151-73.

Delmas P, Brown DA. Pathways modulating neural KCNQ/M (Kv7) potassium channels. *Nat. Rev. Neurosci.* 2005;6:850-62.

Devaux JJ, Kleopa KA, Cooper EC, Scherer SS. KCNQ2 is a nodal K+ channel. *J. Neurosci.* 2004;24:1236-44.

Goodman AD, Brown TR, Edwards KR, Krupp LB, Schapiro RT, Cohen R, et al. A phase 3 trial of extended release oral dalfampridine in multiple sclerosis. *Ann. Neurol.* 2010;68:494-502.

Gunasekaran R, Narayani RS, Vijayalakshmi K, Alladi PA, Shobha K, Nalini A, et al. Exposure to cerebrospinal fluid of sporadic amyotrophic lateral sclerosis patients alters Nav1.6 and Kv1.6 channel expression in rat spinal motor neurons. *Brain Res.* 2009;1255:170-9.

Gutman GA, Chandy KG, Adelman JP, Aiyar J, Bayliss DA, Clapham DE, et al. International Union of Pharmacology. XLI. Compendium of voltage-gated ion channels: potassium channels. *Pharmacol. Rev.* 2003;55:583-6.

Hart IK, Maddison P, Newsom-Davis J, Vincent A, Mills KR. Phenotypic variants of autoimmune peripheral nerve hyperexcitability. *Brain.* 2002;125:1887-95.

Kanai K, Kuwabara S, Misawa S, Tamura N, Ogawara K, Nakata M, et al. Altered axonal excitability properties in amyotrophic lateral sclerosis: impaired potassium channel function related to disease stage. *Brain.* 2006;129:953-62.

Kiernan M, Burke D, Andersen K, Bostock H. Multiple measures of axonal excitability: a new approach in clinical testing. *Muscle Nerve.* 2000;23:399-409.

Kiss T. Persistent Na-channels: origin and function. A review. *Acta Biol Hung.* 2008;59 Suppl:1-12.

Kleine B, Stegeman D, Schelhaas H, Zwarts M. Firing pattern of fasciculations in ALS: evidence for axonal and neuronal origin. *Neurology.* 2008;70:353-9.

Maljevic S, Wuttke TV, Lerche H. Nervous system KV7 disorders: breakdown of a subthreshold brake. *J. Physiol.* 2008;586:1791-801.

Maljevic S, Wuttke TV, Seebohm G, Lerche H. KV7 channelopathies. *Pflugers Arch.* 2010;460:277-88.

Maruyama H, Morino H, Ito H, Izumi Y, Kato H, Watanabe Y, et al. Mutations of optineurin in amyotrophic lateral sclerosis. *Nature.* 2010;465:223-6.

Nodera H, Kaji R. Nerve excitability testing and its clinical application to neuromuscular diseases. *Clin. Neurophysiol.* 2006;117:1902-16.

Pieri M, Carunchio I, Curcio L, Mercuri NB, Zona C. Increased persistent sodium current determines cortical hyperexcitability in a genetic model of amyotrophic lateral sclerosis. *Exp. Neurol.* 2009;215:368-79.

Schwarz J, Glassmeier G, Cooper E, Kao T, Nodera H, Tabuena D, et al. KCNQ channels mediate IKs, a slow K+ current regulating excitability in the rat node of Ranvier. *J. Physiol.* 2006;573:17-34.

Shibuta Y, Nodera H, Nodera A, Okita T, Asanuma K, Izumi Y, et al. Utility of recovery cycle with two conditioning pulses for detection of impaired axonal slow potassium current in ALS. *Clin. Neurophysiol.* 2010;121:2117-20.

Sittl R, Carr RW, Schwarz JR, Grafe P. The Kv7 potassium channel activator flupirtine affects clinical excitability parameters of myelinated axons in isolated rat sural nerve. *J. Peripher Nerv Syst.* 2010;15:63-72.

Tamura N, Kuwabara S, Misawa S, Kanai K, Nakata M, Sawai S, et al. Increased nodal persistent Na+ currents in human neuropathy and motor neuron disease estimated by latent addition. *Clin. Neurophysiol.* 2006;117:2451-8.

Turner BJ, Talbot K. Transgenics, toxicity and therapeutics in rodent models of mutant SOD1-mediated familial ALS. *Prog. Neurobiol.* 2008;85:94-134.

Vucic S, Kiernan M. Axonal excitability properties in amyotrophic lateral sclerosis. *Clin. Neurophysiol.* 2006;117:1458-66.

Vucic S, Kiernan MC. Upregulation of persistent sodium conductances in familial ALS. *J. Neurol Neurosurg. Psychiatry.* 2010;81:222-7.

Wang HS, Pan Z, Shi W, Brown BS, Wymore RS, Cohen IS, et al. KCNQ2 and KCNQ3 potassium channel subunits: molecular correlates of the M-channel. *Science.* 1998;282:1890-3.

Waxman SG. Axonal conduction and injury in multiple sclerosis: the role of sodium channels. *Nat. Rev. Neurosci.* 2006;7:932-41.

Zona C, Pieri M, Carunchio I. Voltage-dependent sodium channels in spinal cord motor neurons display rapid recovery from fast inactivation in a mouse model of amyotrophic lateral sclerosis. *J. Neurophysiol.* 2006;96:3314-22.

In: Potassium Channels
Editor: Danielle S. Fonseca

ISBN: 978-1-61324-880-5

Chapter V

Plant Mitochondrial Potassium: Channel or Channels?*

Donato Pastore*†, *Maura Nicoletta Laus* and *Mario Soccio

Dipartimento di Scienze Agro-ambientali,
Chimica e Difesa Vegetale, Facoltà di Agraria,
Università degli Studi di Foggia,
Centro di Ricerca Interdipartimentale BIOAGROMED,
Università degli Studi di Foggia, Foggia, Italy

Abstract

The history of plant mitochondrial potassium channel/s began about ten years ago, when the first channel was described on a functional basis in durum wheat mitochondria.

This channel was named Plant Mitochondrial Potassium Channel ATP sensitive (PmitoKATP) in analogy with the animal counterpart

---

* A version of this chapter was also published in *Mitochondria Structure, Functions and Dysfunctions* edited by Oliver L. Svensson, published by Nova Science Publishers, Inc. It was submitted for appropriate modifications in an effort to encourage wider dissemination of research.

† Corresponding author: Dipartimento di Scienze Agro-ambientali, Chimica e Difesa Vegetale, Facoltà di Agraria, Università degli Studi di Foggia, Via Napoli, 25 - 71122 Foggia - Italy. tel: +39(0)881589249, fax +39(0)881589342, e-mail: d.pastore@unifg.it

(mitoKATP). The PmitoKATP shows interesting features, being able to deeply affect mitochondrial bioenergetics and to control mitochondrial reactive oxygen species (ROS) production, thus playing a role as a mechanism acting against oxidative/environmental stresses in mitochondria/cell/plant.

To date, mitochondrial potassium channels have been also described in about ten plant species. However, these channels display at the same time analogies and differences with respect to the original PmitoKATP. The intrinsic activity may vary significantly among channels, the pattern of modulators is different and even some of these channels are ATP-insensitive.

Finally, different physiological roles have been proposed. Awaiting for the identification of the molecular nature of each channel, here we point out the question whether the observed differences may be attributed to the same channel differently modulated in mitochondria from different sources or it may be more appropriate to consider different channels. At the moment, it appears more appropriate to refer to different channels, thus suggesting that a fit nomenclature should be consistently used.

Potassium Channels in Cells and Mitochondria

Potassium is the most abundant inorganic cation in the cytosol of both animal and plant cells, where it regulates several important functions such as osmoregulation, control of membrane polarization, electrical neutralisation of dissociated organic acids and anionic groups of macromolecules (nucleic acids and phospholipids) (Lebaudy et al., 2007).

To fulfil these requirements, K+ is readily transported across the plasma membrane mainly by means of K+ channels, multimeric integral membrane proteins forming transmembrane aqueous pores through which K+ specifically permeates. In animals, plasma membrane K+ channels are involved in a variety of important physiological functions of both excitable and nonexcitable cells, including neurotransmitter release, heart rate, insulin secretion, neuronal excitability, epithelial electrolyte transport, smooth muscle contraction, cell volume regulation (Shieh et al., 2000). Several human genetic diseases, such as pathologies involving cardiac arrhythmias, deafness, epilepsy, diabetes, and misregulation of blood pressure, are caused by disruption of K+ channel genes (Miller, 2000).

The existence of K+ transport systems has been demonstrated also in mitochondria. In mammals an inner mitochondrial membrane-located K+

channel has been well characterized, specifically inhibited by ATP (Inoue, 1991; Paucek et al., 1992; Garlid, 1996) and involved in matrix volume homeostasis (Yarov-Yarovoy et al., 1997), regulation of mitochondrial energy metabolism, control of protonmotive force (Czyz et al., 1995), cell signalling and cardioprotection (Mironova et al., 2007; Zingman et al., 2007).

In plants a large number of plasma membrane K+ channels have been described, involved in different tissue-specific and cell-specific processes related to plant growth and development, including germination, root and shoot growth, K+ uptake from soil solution into roots, long-distance K+ transport in the xylem and phloem, K+ fluxes in guard cells during stomatal movements, tropisms, nutrient storage, osmoregulation (Chérel, 2004; Lebaudy et al., 2007).

As for plant mitochondria, to date, K+ channels have been described in about ten plant species (Table 1).

A protein-mediated K+ uptake has been evidenced for the first time about ten years ago in mitochondria isolated from etiolated seedlings of durum wheat (Pastore et al., 1999). In these mitochondria K+ uptake is mediated by an ATP-dependent K+ channel, named PmitoKATP (Pastore et al., 1999; 2007) in analogy with the animal counterpart (mitoKATP, Paucek et al., 1992; Garlid, 1996). Successively, another ATP-sensitive K+ import pathway has been described and well characterized in mitochondria obtained from etiolated pea stem (Petrussa et al., 2001; Chiandussi et al., 2002; Casolo et al., 2003; Petrussa et al., 2004). A mitochondrial ATP-sensitive K+ uptake has been also shown in soybean cell cultures (Casolo et al., 2005). Very recently, the existence of an ATP-sensistive K+ uniporter has been reported in mitochondria from embryogenic cultures of Picea abies (L.) Karst. and Abies cephalonica Loud (Petrussa et al., 2008a) and from Arum spadix (Petrussa et al., 2008b). Moreover, numerous evidences of the existence of a K+ uniport have been also obtained in mitochondria from different plants/organs/tissues (Table 1). In rice seedlings, durum wheat roots and Arabidopsis cell cultures, the ATP-sensitivity of the mitochondrial K+ transport has also been investigated and evidenced (Pastore et al., unpublished data).

On the other hand, an ATP-insensitive quinine-inhibited K+ transporter has been described in mitochondria from potato tubers, etiolated seedlings of maize and tomato fruits (Ruy et al., 2004). In mitochondria from a cell culture of potato, a K+ transport was observed, but no investigation was carried out about its ATP-sensitivity (Fratianni et al., 2001). Moreover, in potato tuber mitochondria a large-conductance calcium-activated K+ channel has been also recently described, but no information is provided regarding its ATP-sensitivity (Koszela-Piotrowska et al., 2009).

Table 1. Plant species in which the existence of a mitochondrial K+ channel has been shown

Plant species	tissue/organ	sensitivity to ATP
Arum italicum Mill. and Arum maculatum L.[a, b]	Spadix (Petrussa et al., 2008[b])c	sensitive
Durum wheat (Triticum durum Desf.)[a,b]	Etiolated seedlings (Pastore et al., 1999)	sensitive
Greek fir (Abies cephalonica Loud) [a, b]	Embryogenic cultures (Petrussa et al., 2008a)	sensitive
Maize (Zea mays L.)[b]	Etiolated seedlings (Ruy et al., 2004)	insensitive
Norway spruce (Picea abies (L.) Karst.) [a, b]	Embryogenic cultures (Petrussa et al., [2008a])	sensitive
Pea (Pisum sativum L.)[a,b]	Etiolated stem (Petrussa et al., 2001, 2004) (Chiandussi et al., 2002) (Casolo et al., 2003)	sensitive
Potato (Solanum tuberosum L.)[a,b]	Tuber (Ruy et al., 2004) (Koszela-Priotrowska et al., 2009)	insensitive $Ca2^{+}$-dependent
Soybean (Glycine max L.)[a,b]	Suspension cell cultures (Casolo et al., 2005)	sensitive
Tomato (Lycopersicum esculentum Mill.)[b]	Green fruit (Ruy et al., 2004)	insensitive

Plants/organs/tissues in which a mitochondrial K+ influx has been shown (Pastore et al., 1999; 2001; unpublished data).

Etiolated seedlings: bread wheat (Triticum aestivum L.)a,b; barley (Hordeum vulgare L.)[a,b]; lentil (Lens esculenta Moench.)b; rice (Oryza sativa L.)[a,b]; rye (Secale cereale L.)[a,b;] spelt (Triticum dicoccum Schübler)[a,b]; triticale (Triticum x Secale)[b]

[a,b] Green leaves: durum wheat (Triticum durum Desf.)a,b; maize (Zea mays L.)b; spinach (Spinacea oleracea L.).

Roots: durum wheat (Triticum durum Desf.)[a,b]; pea (Pisum sativum L.)[a,b].

Suspension cell cultures: potato (Solanum tuberosum L.)[a,b]; Arabidopsis thaliana (L.) Heynha,[b.]

Tubers: potato (Solanum tuberosum L.)[a,b]; topinambur (Helianthus tuberosus L.)[b].

[a] Activity evaluated by experiments, b Activity evaluated by swelling experiments (see Pastore et al., 1999), [c] Bibliographic reference.

Variety of Properties and Functions of Plant Mitochondrial K+ Channels

As reported above, a major property of the K+ channel(s) described in different plant (tissues/organs) species is the ATP-regulation. However, the well characterized ATP-dependent K+ channels also display some different biochemical properties, thus suggesting that we should refer to different mitochondrial K+ channels.

In durum wheat mitochondria (DWM) PmitoKATP is inhibited by ATP [Ki = 290 μM, about tenfold lower than that of the mammalian counterpart (Garlid, 1996)], NADH (Ki = 390 μM) and Zn2+ and activated by GTP and diazoxide, that counteract ATP inhibition, and by palmitoylCoA, CoA and thiol reagents; Mg2+ and glyburide, which inhibit the mammalian channel, are ineffective in DWM. K+ uptake by DWM, evaluated by means of experiments in which electrical transmembrane potential ($\Delta\Psi$) changes are fluorimetrically monitored, shows hyperbolic dependence on K+ concentration (Vmax=12.5 ± 1.96 mV/s, Km=2.2 ± 0.78 mM). Moreover, K+ influx, evaluated by means of swelling experiments, exhibits monovalent cation specificity in the following order Cs+>K+=Rb+>Na+=Li+ (Pastore et al., 1999; 2007); it is notable that Cs+, that usually inhibits plant K+ channels (Ichida and Schroeder, 1996), is transported by PmitoKATP with an efficiency even higher with respect to K+. PmitoKATP shows a $\Delta\Psi$-dependence: the channel activity remains rather constant in the range 95-140 mV, while it linearly increases in $\Delta\Psi$ range varying from 140 to 175 mV. PmitoKATP is also activated as a consequence of mitochondria treatment with the superoxide anion producing system, xanthine plus xanthine oxidase. The PmitoKATP-mediated electrophoretic uniport of K+ across the inner mitochondrial membrane towards the matrix is able to completely collapse the electrical membrane potential, $\Delta\Psi$ [the main component of protonmotive force in plant mitochondria (Douce et al., 1987)] (Pastore et al., 1999; 2007).

The ATP-sensitive K+ channel described in etiolated pea stem mitochondria differs from PmitoKATP in some characteristics, first of all the activation by cyclosporin A (CsA). Moreover, whereas in DWM a K+ efflux outside mitochondria has never been observed also in the presence of the selective ionophore valinomycin, in de-energized pea mitochondria, CsA induced an outwardly directed K+ diffusion potential, as suggested by $\Delta\Psi$ generation due to CsA addition to mitochondria, further increased by valinomycin (Petrussa et al., 2001); interestingly, CsA-induced K+ efflux was inhibited by increasing KCl concentrations, with a sigmoidal kinetic, that became hyperbolic in mitochondria treated with CsA after 2 h from extraction, but that was maintained even after 2 h in mitochondria treated with 5 mM DTE (Petrussa et al., 2001). Moreover, CsA-induced K+ diffusion potential was prevented by monovalent cations in the following order: K+>Rb+=Li+>Na+. At the same time, CsA also induced a K+ influx, since it stimulated mitochondrial swelling in KCl medium and favoured a more rapid KCl-dependent depolarization in succinate-energized mitochondria; the $\Delta\Psi$ dissipation by KCl addition to purified mitochondria was only partial and smaller than that observed in washed mitochondria. Briefly, unlike PmitoKATP, in pea steam mitochondria K+ channel catalyzes both K+ influx and efflux under the adopted experimental conditions, and shows a high selectivity for K+ as well as a remarkable ability to discriminate between K+ and Rb+; moreover, it is voltage-dependent, but it tends to close with $\Delta\Psi$ increasing and appears to be modulated by redox state of mitochondria, possibly through a dithiol-disulfide interconversion. Furthermore, the KATP channel of pea stem mitochondria resulted activated by GTP and diazoxide, that partially restored ATP-inhibition, thiol reagents (phenylarsine oxide and N-ethyl maleimide, NEM) and nitric oxide (NO), and inhibited by glyburide, 5-hydroxydecanoate and hydrogen peroxide (H_2O_2) (Chiandussi et al., 2002).

K+ channel(s) described in potato, maize and tomato mitochondria display properties strongly different from those of durum wheat and pea mitochondria K+ channels. In these mitochondria a highly active K+ uptake not inhibited by ATP was described (Ruy et al., 2004). A partial inhibitory effect of ATP was observed on potato mitochondrial swelling, but prevented by BSA (that removes fatty acids) and the H+ ionophore FCCP; in the light of this, the authors invoked the occurrence of an ATP-insensitive K+ transporter and an ATP-inhibited K+ pathway involving K+ uptake through the K+/H+ antiporter in conjunction with an uncoupling protein (PUCP)-mediated H+ cycling. The same behaviour was observed in maize and tomato mitochondria. Externally added K+ to succinate-respiring potato mitochondria was found to induce a large increase in oxygen consumption, exhibiting an hyperbolic dependence with ion concentration (K1/2=27.9 ± 1.6 mM), and not affected by ATP, NADH, glyburide and 5-hydroxydecanoate, but inhibited by quinine, a broad spectrum K+ channel inhibitor, in a dose-dependent manner (K1/2=254 ± 10μM).

Plant K+ channels have been proposed to perform different functional roles. In durum wheat it has been demonstrated that PmitoKATP-induced $\Delta\Psi$ dissipation is able to lower mitochondrial reactive oxygen species (ROS) production (Pastore et al., 1999). So, PmitoKATP is suggested to cooperate with antioxidant cellular systems and with the other mitochondrial energy-dissipating systems, the PUCP (Pastore et al., 2000) and the alternative oxidase (Pastore et al., 2001), to defend the cell from oxidative stress occurring when plants suffer environmental stresses (Pastore et al., 2007); this was shown in durum wheat seedlings suffering hyperosmotic stress (Trono et al., 2004) as well as in water stress adapted potato cells (Fratianni et al., 2001).

In pea stem mitochondria K+ channel functioning has been reported to play a crucial role in controlling mitochondrial succinate-dependent H2O2 production (Casolo et al., 2003). Moreover, the aperture of this channel is known to be linked to the partial rupture of the outer membrane, with a consequent release of cytochrome c (Chiandussi et al., 2002), and to be involved in low-amplitude permeability transition induction (Petrussa et al., 2004), typical events of early phases of induced programmed cell death (Zottini et al., 2002). The involvement of K+ channel also in programmed cell death manifestation and regulation has been demonstrated in soybean cell cultures treated with H_2O_2 or NO (Casolo et al., 2005). In Arum spadix mitochondria the ATP-sensitive K+ channel is able to induce mitochondrial uncoupling associated to thermogenesis induction (Petrussa et al., 2008b); in mitochondria from embryogenic cultures of Picea abies (L.) Karst. and Abies cephalonica Loud the CsA-sensitive KATP channel may be probably involved in stress response (Petrussa et al., 2008a). Very recently, a role of KATP channel in inducing mitochondrial swelling and cytochrome c release during the proliferation stage of programmed cell death manifestation in Abies alba somatic embryogenesis has been suggested (Petrussa et al., 2009).

The ATP-insensitive quinine-inhibited K+ pathway described in potato, tomato and maize mitochondria is suggested to be involved in metabolism regulation and prevention of ROS formation (Ruy et al., 2004).

Variety of Possible Molecular Identities of Plant Mitochondrial K+ Channels

To date, the molecular identity of PmitoKATP of durum wheat is unknown, as well as a very few information is available about gene encoding other plant mitochondrial K+ channel(s).

As regards the animal counterpart, nonetheless the existence of an ATP-sensitive K+ channel in mammalian mitochondria is known since 1991 (Inoue et al., 1991) and its central role in myocardial protection and preservation of cardiac function following an ischemic insult has been extensively studied (Zingman et al., 2007), to date, information on its molecular identity is rather controversial. Three different research teams have reported three different results. Garlid and coworkers isolated, purified and reconstituted in liposomes a KATP channel from rat liver, beef heart (Paucek et al., 1992) and rat brain (Bajgar et al., 2001) mitochondria. This channel (mitoKATP) (Figure 1A), like its counterpart in the plasma membrane, proved to be a hetero-multimeric complex consisting of four 55-kDa inward rectifying K+ subunits (mitoKir) [1 P/2 TMS type α subunits, each containing two transmembrane segments (TMS) with a pore-forming region (P-loop) in between (Miller, 2000; Shieh et al., 2000)] and four 63-kDa sulfonylurea-binding regulatory subunits (mitoSUR) [members of the ATP-binding cassette (ABC) protein superfamily (Shieh et al., 2000)] (Grover and Garlid, 2000; Mironova et al., 2004; Carreira et al., 2005). On the other hand, another research group has demonstrated the localization in the inner mitochondrial membrane of T lymphocytes (Szabò et al., 2005) and, recently, of human colon cancer cells (De Marchi et al., 2009) of the 1 P/6 TMS type [consisting of six TMS (S1-S6) with a voltage sensor located at S4 and a P loop between S5 and S6] K+ channels Kv1.3 [outward rectifying K+ channels arranged as octameric complex, composed of four α subunits and four auxiliary regulatory β subunits (Kvβ) (Miller, 2000; Shieh et al., 2000)] and KCa3.1 (Ca2+-activated), respectively, traditionally thought to be exclusively localized in plasma membrane (Figure 1B). Finally, Ardehali et al. (2004) have found mitoKATP channel activity in a proteoliposome-reconstituted highly purified fraction of the inner mitochondrial membrane containing five different mitochondrial proteins, mitochondrial ATP-binding cassette protein 1 (mABC1), phosphate carrier, adenine nucleotide translocator, ATP sinthase and succinate dehydrogenase, but lacking Kir or Kv type subunits (Figure 1C). These results clearly show that in mammal mitochondria we may refer to different potassium channels.

In plants the molecular identification of K+ channel genes is still at the beginning; at present most of knowledge refers to Arabidopsis thaliana. Fifteen genes have been identified in Arabidopsis belonging to three distinct K+ channel families. The 1 P/6 TMS type K+ channel family includes nine genes homologous to the animal Shaker-type Kv genes and it is subdivided in two branches: the former consists of AKTs and KATs genes, encoding inward rectifying K+ channels; the latter includes the outward rectifying SKOR and GORK (Mäser et al., 2001; Lebaudy et al., 2007).

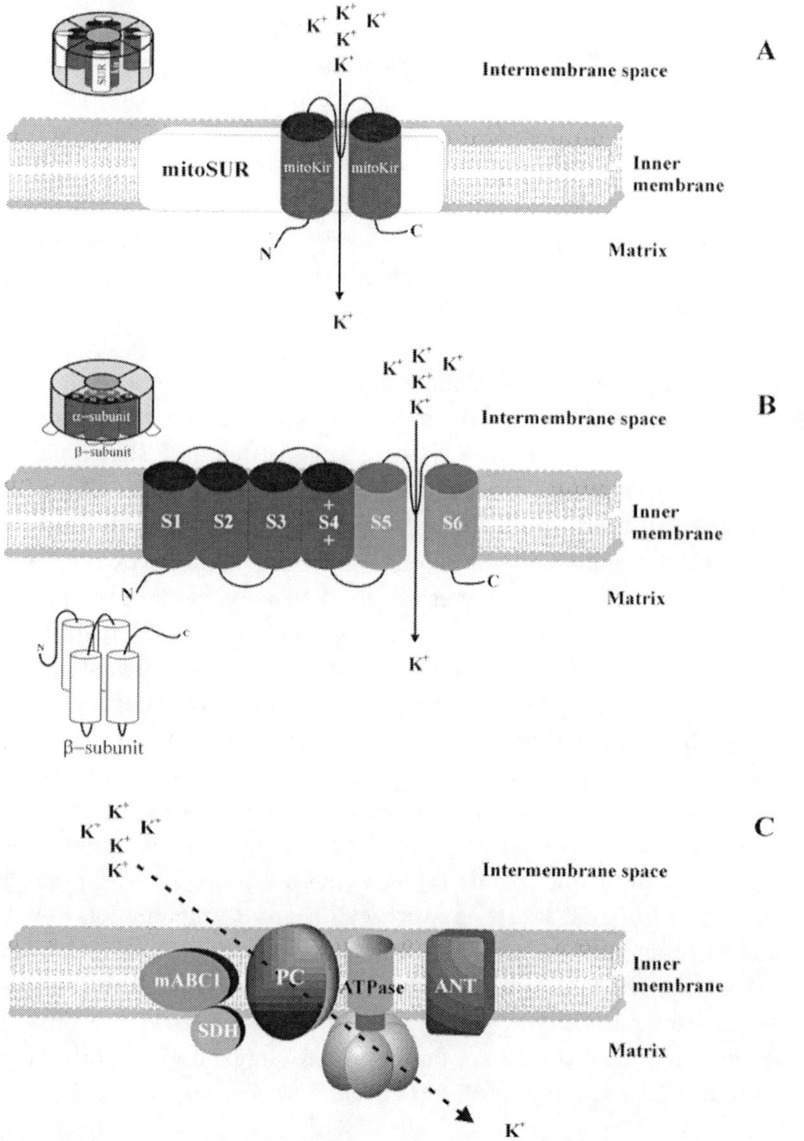

Figure 1. Structures proposed for the mitochondrial K+ channel(s) in mammals. Abbreviations: S, transmembrane segments; mABC1, mitochondrial ATP-binding cassette protein 1; ANT, adenine nucleotide translocator; SDH, succinate dehydrogenase; PC, phosphate carrier. For the explanation see the text. Revised from Mironova et al. (2004) (A), A SIGMA catalogue (B), Ardehali and O'Rourke (2005) (C).

The 2P/4 TMS family includes five genes encoding “two pore or tandem-pore” outward rectifying K+ channels, which are dimers of α subunits, each consisting of four TMS and two P loops (Mäser et al., 2001; Chérel, 2004; Lebaudy et al., 2007). On the contrary, the 1 P/2 TMS Kir-like channel family is represented by the only KCO3 gene (Czempinski et al., 1999; Lebaudy et al., 2007). In carrot the expression of KDC1 gene, coding for an inward rectifying shaker K+ channel, has been described and characterized in early stages of somatic embryonic development (Downey et al., 2000; Costa et al., 2004); a shaker K+ channel (DKT1) expressed during later stages of the same process has been also reported (Formentin et al., 2004).

As concerns plant mitochondrial K+ channel(s), to date very few information is available about the molecular identity. In a congress report, the partial purification of a mitoKATP, consisting of Kir and SUR type subunits, from potato was proposed (Paucek et al., 2002).

On the other hand, by means of proteomic approach, the identification of a K+ channel β regulatory subunit has been reported in rice mitochondria (Tanaka et al., 2004), thus suggesting the existence of Kv channels in plant mitochondria. Very recently, by means of electrophysiological studies and immunological analysis, a large-conductance calcium-activated K+ channel similar to that of mammalian mitochondria [mitoBKCa channel, (Siemen et al., 1999)] has been described in potato tuber, consisting of a pore-forming subunit and an auxiliary 2 subunit (Koszela-Piotrowska et al., 2009). However, it should be outlined that data obtained on the basis of immunoblot analysis should be considered not conclusive. Concerning this, in a recent study, it is reported that two distinct bands (putative Kir6.1 proteins) of 51 and 48 kDa revealed in isolated heart mitochondria by immunoblots with two commercially available antibodies, after purification were identified by LC-MS/MS as NADH-dehydrogenase flavoprotein 1 and mitochondrial NADP+-dependent isocitrate dehydrogenase, respectively (Foster et al., 2008). In the whole, although these results are only initial findings, they suggest that also in plant mitochondria we might refer to different potassium channels.

The Relationship between Properties/Functions and Molecular Identities

It should be underlined that it is not necessary to invoke a different molecular identity to justify the different functional properties of plant mitochondrial potassium channels described by different authors. Consistently, although many different types of K+ channels have been evidenced in plants, they are encoded by only a relative small number of genes. So, Reintanz et al. (2002) pointed out the question how the diversity of K+ currents can be created by such a small numbers of proteins.

The molecular basis for so high functional diversity of plant K+ channels was suggested by the authors to be provided, besides by multiple genes, also by homo- and hetero-multimeric assembly of identical and related α subunits and their association with regulatory subunits and by regulation of channel genes and proteins at the post-transcriptional and/or at the post-translational level, respectively (Reintanz et al., 2002 and refs. therein).

Concerning this question, here we report some interesting and intriguing evidences gained in DWM relative to ATP-sensitivity and activity of PmitoKATP. In Figure 2 two different experiments aimed to evaluate the ATP-sensitivity of PmitoKATP were reported, in which $\Delta\Psi$ changes were measured by measuring the fluorimetric probe safranin "O" (Pastore et al., 1996). In the experiment of Figure 2A, 25 mM KCl addition to succinate-energized DWM (state 4 of respiration, i.e. in the absence of ADP) induced only a slow $\Delta\Psi$ decrease; 10 M diazoxide, a potassium channel opener, caused a faster and almost complete depolarization, and 0.2 mM ATP completely restored.

On the contrary, in the experiment of Figure 2B, KCl addition completely depolarized DWM, while 1 mM ATP induced only a partial recovery. These are two extreme situations relative to mitochondria obtained from different not homogeneous seed stocks, but in general, in different mitochondrial preparations PmitoKATP may show some variability about activity and sensitivity to ATP: in DWM preparations showing a less active channel, K+ uptake may be strongly inhibited by lower ATP concentrations, while a large conductance channel, i.e. able to completely depolarize DWM at high rate, appears to be only slightly sensitive to ATP inhibition and only at higher inhibitor concentrations.

These results suggest the occurrence of a regulation of the channel protein, although the existence in DWM of multiple channel identities with different K+ conductance and ATP-sensitivity cannot be completely excluded.

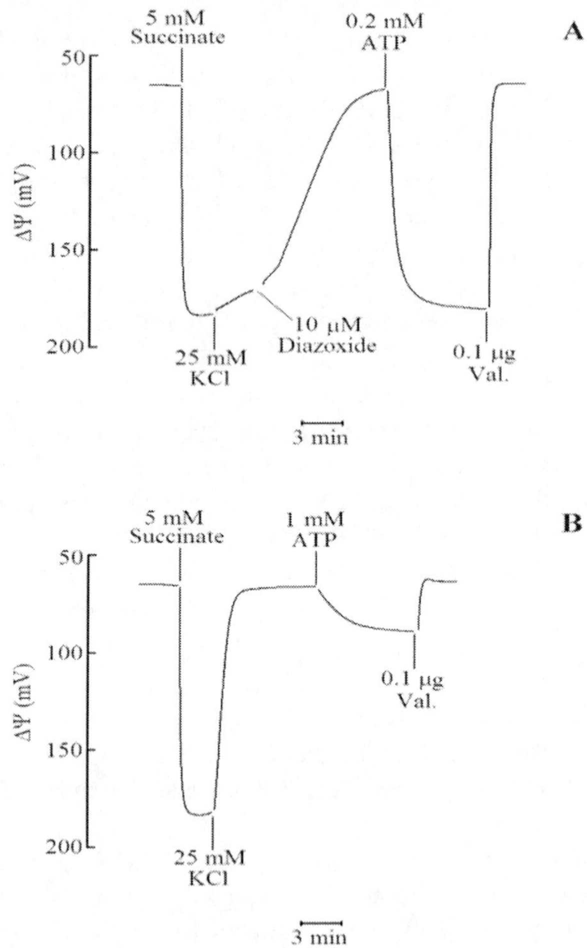

Figure 2. Effect of ATP and Diazoxide on decrease induced by PmitoKATP functioning in DWM. ΔΨ measurements were carried out by monitoring safranin "O" fluorescence changes as a function of time as reported in Pastore et al. (1999); calibration of fluorescence changes vs ΔΨ was carried out as in Pastore et al. (1996). DWM (0.2 mg prot.) were suspended in 2.0 ml of a medium consisting of 0.3 M mannitol, 20 mM Tris-HCl pH 7.20, 5 mM MgCl2, 2.5 μM safranin. Additions of succinate, KCl, diazoxide, ATP and valinomycin (Val.) at the reported concentrations were carried out at the time indicated.

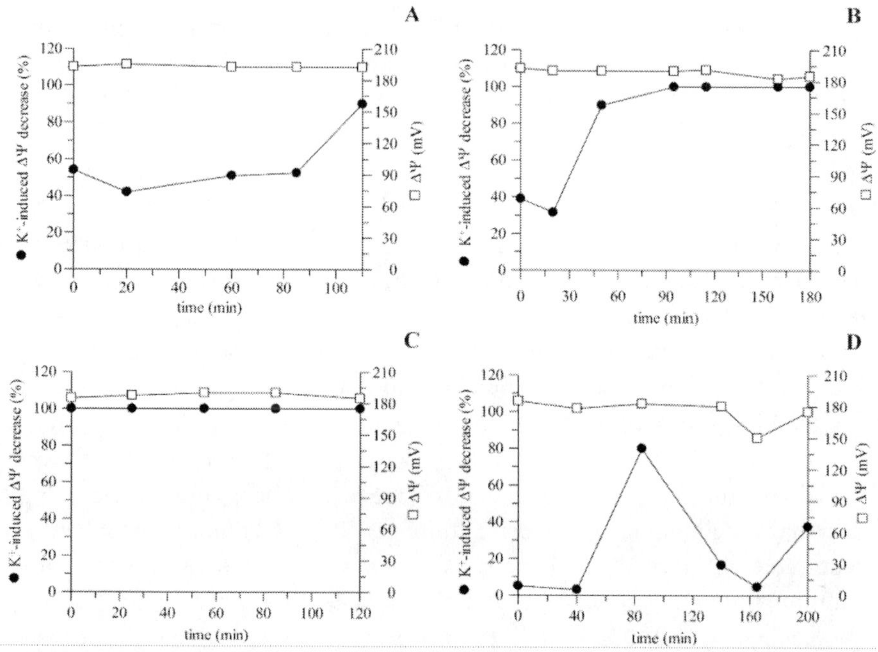

Figure 3. Timing of membrane potential (ΔΨ) generated by succinate oxidation and ΔΨ decrease induced by PmitoKATP functioning in different DWM preparations. ΔΨ measurements were carried out as described in Figure 2. Succinate (5 mM) was used as a respiratory substrate to generate membrane potential (ΔΨ); then, 25 mM KCl was added to evaluated the depolarization due to PmitoKATP. ΔΨ values and the K+-induced ΔΨ decreases (expressed as percent of ΔΨ values) are reported, as obtained in successive measurements carried out in the course of four different experiments (A-D).

In the experiments of Figure 3ΔΨ the value of generated by succinate oxidation was evaluated in four different DWM preparations, together with the decrease induced by KCl. As in Figure 2, 5 mM succinate was added to generate ΔΨ and 25 mM KCl was added to evaluate the depolarization induced by PmitoKATP opening. In x-axis, the timing from the beginning of each experiment is reported. Always, a high ΔΨ (about 180 mV) was measured, that remained significantly unchanged in the course of the experiment, thus indicating the maintenance of good membrane intactness and functionality. On the contrary, a different activity of PmitoKATP was observed in the different preparations: in a DWM preparation (A), a partially active channel that tended to completely open only at the end of the experiment was observed; in another mitochondrial preparation (B), a full opening of PmitoKATP early occurred, about 45 min after the beginning of the experiment; in (C) a completely opened PmitoKATP was observed, with a full activity maintained during the entire course of the experiment; finally, in (D) the DWM preparation was characterized by a repeatedly changing PmitoKATP activity.

In another experiment (Figure 4) ΔΨ decrease induced by PmitoKATP functioning was evaluated in the same DWM preparation together with the succinate-supported ΔΨ generation, as well as respiratory control (RC) and ADP/O ratios (these last two parameters are relative to the oxidation of the malate plus glutamate substrate pair). RC and ADP/O ratios, providing an estimation of the mitochondrial coupling between oxidation and phosphorylation processes and the phosphorilative efficiency, respectively, are also good indicators of mitochondrial intactness and functionality. A high ΔΨ value of about 170 mV (A) and good RC (B) and ADP/O ratio (C) values were maintained during the entire course of the experiment, but in these highly intact and functional mitochondria, PmitoKATP activity largely changed: from a complete inactive state the channel rapidly passed to a full functioning condition, but then it slowly returned to a completely inactive state.

Succinate (5 mM) and KCl (25 mM) were added to induce ΔΨ generation and PmitoKATP opening-dependent depolarization, respectively. In A, ΔΨ values and the K+-induced ΔΨ decreases (expressed as percent of ΔΨ values) measured in the course of the experiment are reported. In the same experiment oxygen uptake rate was measured as described in Pastore et al. (2001). As respiratory substrate pair, malate (10 mM) plus glutamate (2 mM, in order to allow oxaloacetate removal via transamination) in the presence of 200 μM thiamine pyrophosphate (in order to activate the 2-oxoglutarate dehydrogenase complex) were added. In the course of substrate oxidation, successive additions of 100 μM ADP were carried out in order to evaluate: (i) the respiratory control (RC) ratio, i.e. the ability of ADP to control the oxygen uptake rate expressed as a ratio between the rate measured in the presence of ADP plus Pi (state 3) and the rate measured after ADP consumption (state 4); (ii) the ADP/O ratio, i.e. the ratio between the nmol of phosphorylated ADP and the natom of oxygen consumed.

In B and C the RC and ADP/O ratio values measured in the course of the experiment are reported, respectively, together with the K+-induced ΔΨ decreases. RC and ADP/O ratio values are reported as mean value ± standard deviation (n= 3, since 3 successive ADP additions were carried out in each measurement).

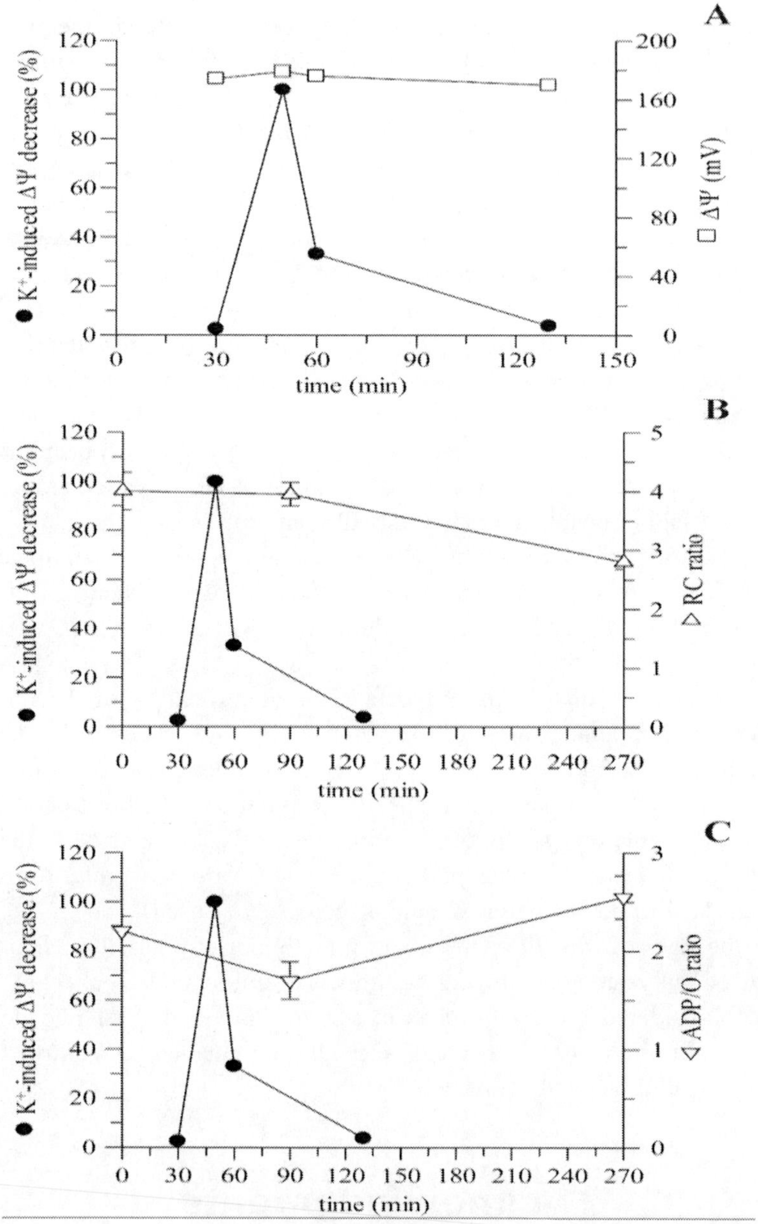

Figure 4. Timing of membrane potential (ΔΨ) generated by succinate oxidation, ΔΨ decrease induced by PmitoKATP functioning and respiratory control (RC) and ADP/O ratios in the same DWM preparation. ΔΨ measurements were carried out as described in Figure 3.

So, experiments in Figures 2-4 show that, in DWM, the activation-inactivation (opening-closing?) of PmitoKATP, occurring with a still unknown mechanism not related to mitochondrial integrity and coupling, may be observed not only in different experiments, but also, during the time, in the same mitochondrial preparation, thus suggesting that a conversion of the same K+ channel protein from an active into an inactive state and vice versa is possible. As regards this point, a phosphorylation/dephosphorylation mechanism has been already suggested for the regulation of Arabidospis K+ transporter 1 (AKT1) activity in roots. AKT1 channel, in fact, is activated through phosphorylation by CIPK23 [CBL(calcineurin B-like calcium sensor)-interacting protein kinase 23] and dephosphorylated and inactivated due to interaction with AIP1 (AKT1-interacting 2C-type protein phosphatase) (Lee et al., 2007). It should be outlined that a phosphorylation/dephosphorylation network is also responsible of mammalian mitoKATP activation associated to mitochondrial permeability transition (MPT) inhibition: K+ channel is opened by the external protein kinase G(PKG) via the endogenous mitochondrial PKCε1and the consequent K+ influx leads to an increase in matrix H2O2, that in turn activates PKCε2 causing inhibition of MPT opening (Costa et al., 2006).

In conclusion, in the absence of a clear identification of the molecular nature of each channel, the question whether the differences in K+ transport observed in plant mitochondria from different sources may be attributed to a species-specific or tissue(organ)-specific regulation of the same channel protein or to the existence of different channels, remains still unsolved. However, at the moment, it is probably more appropriate to refer to different channels, so a fit nomenclature should be consistently used.

In the light of this, the name Plant mitochondrial potassium channel, we introduced ten years ago studying durum wheat mitochondrial channel, should be now considered too much generic; so, the PmitoKATP should be better named as Durum Wheat mitochondrial potassium channel, DWmitoKATP, to avoid misleading generalization.

Acknowledgments

This work was supported by the projects MIUR “AGROGEN”.

References

Ardehali, H; Chen, Z; Ko, Y; Mejia-Alvarez, R; Marbán, E. Multiprotein complex containing succinate dehydrogenase confers mitochondrial ATP-sensitive K+ channel activity. *Proceedings of the National Academy of Science of USA*, 2004, 101, 11880-11885.

Ardehali, H; O'Rourke, B. Mitochondrial KATP channels in cell survival and death. *Journal of Molecular and Cellular Cardiology*, 2005, 39, 7-16.

Bajgar, R; Seetharaman, S; Kowaltowski, AJ; Garlid, KD; Paucek, P. Identification and properties of a novel intracellular (mitochondrial) ATP-sensitive potassium channel in brain. *The Journal of Biological Chemistry,* 2001, 276, 33369-33374.

Cao, Y; Ward, JM; Kelly, WB; Ichida, AM; Gaber RF; Anderson, JA; Uozumi, N; Schroeder, JI; Crawford, NM. Multiple genes, tissue specificity, and expression-dependent modulation contribute to the functional diversity of potassium channels in Arabidopsis thaliana. *Plant Physiology*, 1995, 109, 1093-1106.

Carreira, RS, Facundo, HTF; Kowaltowski, AJ. Mitochondrial K+ transport and cardiac protection during ischemia/reperfusion. *Brazilian Journal of Medical and Biological Research*, 2005, 38, 345-352.

Casolo, V; Braidot, E; Chiandussi, E; Vianello, A; Macrì, F. K+ATP channel opening prevents succinate-dependent H2O2 generation by plant mitochondria. *Physiologia Plantarum*, 2003, 118, 313-318.

Casolo, V; Petrussa, E; Krajňáková, J; Macrì, F; Vianello, A. Involvement of the mitochondrial K+ATP channel in H2O2- or NO-induced programmed death of soybean suspension cultures. *Journal of Experimental Botany*, 2005, 56, 997-1006.

Chérel I. Regulation of K+ channel activities in plants: from physiological to molecular aspects. *Journal of Experimental Botany*, 2004, 55, 337-351.

Chérel, I. Regulation of K+ channel activities in plants: from physiological to molecular aspects. *Journal of Experimental Botany*, 2004, 55, 337-351.

Chiandussi, E; Petrussa, E; Macrì, F; Vianello, A. Modulation of a plant mitochondrial K+ATP channel and its involvement in cytochrome c release. *Journal of Bioenergetics and Biomembranes*, 2002, 34, 177-184.

Costa, A; Carpaneto, A; Varotto, S., ; Formentin, E; Marin, O; Barizza, E; Terzi, M; Gambale, F; Lo Schiavo, F. Potassium and carrot embryogenesis: Are K+ channels necessary for development? *Plant Molecular Biology,* 2004, 54, 837-852.

Costa, ADT; Jakob, R; Costa, CL; Andrukhiv K; West, IC; Garlid, KD. The mechanism by which the mitochondrial ATP-sensitive K+ channel opening and H2O2 inhibit the mitochondrial permeability transition. *The Journal of Biological Chemistry*, 2006, 281, 20801-20808.

Czempinski, K; Gaedeke, N; Zimmermann, S; Müller-Röber, B. Molecular mechanisms and regulation of plant ion channels. *Journal of Experimental Botany,* 1999 50, 955-966.

Czyz, A; Szewczyk, A; Nalecz, MJ; Wojtczak, L. The role of mitochondrial potassium fluxes in controlling the protonmotive force in energized mitochondria. *Biochemical and Biophysical Research Communications*, 1995, 210, 98-104.

De Marchi, U; Sass,i N; Fioretti, B; Catacuzzeno, L; Cereghetti, GM; Szabò, I; Zoratti, M. Intermediate conductance Ca2+-activated potassium channel (KCa3.1) in the inner mitochondrial membrane of human colon cancer cells. *Cell Calcium*, 2009, 45, 509-516.

Downey, P; Szabò, I; Ivashikina, N; Negro, A; Guzzo, F; Ache, P; Hedrich, R; Terzi, M; Lo Schiavo F. KDC1, a novel carrot root hair K+ channel: cloning characterization, and expression in mammalian cell. *The Journal of Biological Chemistry*, 2000, 275, 39420-39426.

Formentin, E; Varotto, S; Costa, A; Downey, P; Bregante, M; Naso, A; Picco, C; Gambale, F; Lo Schiavo, F. DKT1, a novel K+ channel from carrot, forms functional heteromeric channels with KDC1. *FEBS Letters*, 2004, 573, 61-67.

Foster, DB; Rucker, JJ; Marbán, E. Is Kir6.1 a subunit of mitoKATP? *Biochemical and Biophysical Research Communications*, 2008, 366, 649-656.

Fratianni, A; Pastore, D; Pallotta, ML; Chiatante, D; Passarella, S. Increase of membrane permeability of mitochondria isolated from water stress adapted potato cells. *Bioscience Reports*, 2001, 21, 81-91.

Garlid, KD. Cation transport in mitochondria – the potassium cycle. *Biochimica et Biophysica Acta*, 1996, 1275, 123-126.

Grover, GJ; Garlid, KD. ATP-sensitive potassium channels: A review of their cardioprotective pharmacology. *Journal of Molecular and Cellular Cardiology,* 2000, 32, 677-695.

Ichida, AM; Schroeder, JI. Increased resistance to extracellular cation block by mutation of the pore domain of the Arabidopsis inward-rectifying K+ channel KAT1. *Journal of Membrane Biology*, 1996, 151, 53-62.

Inoue, I; Nagase, H; Kishi, K; Higuti, T. ATP-sensitive K+ channel in the mitochondrial inner membrane. *Nature*, 1991, 352, 244-247.

Koszela-Piotrowska, I; Matkovic, K; Szewczyk, A; Jarmuszkiewicz, W. A large-conductance calcium-activated potassium channel in potato tuber mitochondria. *Biochemical Journal*, 2009, doi:10.1042/BJ20090991.

Lebaudy, A; Véry, A-A; Sentenac, H. K+ channel activity in plants: genes, regulations and functions, *FEBS Letters*, 2007, 581, 2357-2366.

Lee, SC; Lan, WZ; Kim, BG; Li, L; Yong, CH; Pandey, GK; Lu, G; Buchanan, BB; Luan, S. A protein phosphorylation/dephosphorylation network regulates a plant potassium channel. *Proceedings of the National Academy of Sciences of USA,* 2007, 104, 15959-15964.

Mäser, P; Thomine, S; Schroeder JI; Ward, JM; Hirschi, K; Sze, H; Talke, IN; Amtmann, A; Maathuis, FJM; Sanders, D; Harper, JF; Tchieu, J; Gribskov, M; Persans, MW; Salt, DE; Kim, SA; Guerinot, ML. Phylogenetic relationships within cation transporter families of Arabidopsis. *Plant Physiology*, 2001, 126, 1646-1667.

Miller, C. (2000). Genome Biology, http://genomebiology.com/content/pdf/gb-2000-1-4-reviews0004.pdf.

Mironova, GD; Kachayeva, YV; Krylova, IB; Rodionova OM; Balina, MI; Yevdokimova, NR; Sapronov, NS. Mitochondrial ATP-dependent potassium channel. 2. *The role of the channel in protection of the heart against ischemia.* Vestnik Rossiiskoi Akademii Meditsinskikh Nauk (Bulletin of the Russian Academy of Medical Sciences), 2007, 2, 44-50.

Mironova, GD; Negoda, AE; Marinov, BS; Paucek, P; Costa, AD; Grigoriev, SM; Skarga, YY; Garlid, KD. Functional distinctions between the mitochondrial ATP-dependent K+ channel (mitoKATP) and its inward rectifier subunit (mitoKIR). *The Journal of Biological Chemistry*, 2004, 279, 32562-32568.

Pastore, D; Di Martino, C; Bosco, G; Passarella, S. Stimulation of ATP synthesis via oxidative phosphorylation wheat mitochondria irradiated with helium-neon laser. *Biochemistry and Molecular Biology International*, 1996, 39, 149-157.

Pastore, D; Fratianni, A; Di Pede, S; Passarella, S. Effect of fatty acids, nucleotides and reactive oxygen species on durum wheat mitochondria. *FEBS Letters*, 2000, 470, 88-92.

Pastore, D; Stoppelli, MC; Di Fonzo, N; Passarella, S. The existence of the K+ channel in plant mitochondria. *The Journal of Biological Chemistry*, 1999, 274, 26683-26690.

Pastore, D; Trono, D; Laus, MN; Di Fonzo, N; Flagella, Z. Possible plant mitochondria involvement in cell adaptation to drought stress. A case study: durum wheat mitochondria, *The Journal of Experimental Botany*, 2007, 58, 195-210.

Pastore, D; Trono, D; Laus, MN; Di Fonzo, N; Passarella, S. Alternative oxidase in durum wheat mitochondria. Activation by pyruvate, hydroxypyruvate and glyoxylate and physiological role. *Plant and Cell Physiology*, 2001, 42, 1373-1382.

Paucek, P; Mironova, G; Mahdi, F; Beavis, AD; Woldegiorgis, G; Garlid, KD. Reconstitution and partial purification of the glibenclamide-sensitive, ATP-dependent K+ channel from rat liver and beef heart mitochondria. *The Journal of Biological Chemistry*, 1992, 267, 26062-26069.

Petrussa, E; Bertolini, A; Casolo, V; Krajňáková, J; Macrì, F; Vianello, A. Mitochondrial bioenergetics linked to the manifestation of programmed cell death during somatic embryogenesis of Abies alba. *Planta*, 2009, doi: 10.1007/s00425-009-1028-x.

Petrussa, E; Bertolini, A; Krajňáková, J; Casolo, V; Macrì, F; Vianello, A. Isolation of mitochondria from embryogenic cultures of Picea abies (L.) Karst. and Abies cephalonica Loud.: Characterization of a K+ATP channel. *Plant Cell Reports*, 2008a , 27, 137-146.

Petrussa, E; Casolo, V; Peresson, C; Braidot, E; Vianello, A; Macrì, F. The K+ATP channel is involved in a low-amplitude permeability transition in plant mitochondria. *Mitochondrion*, 2004, 3, 297-307.

Petrussa, E; Casolo, V; Peresson, C; Krajňáková, J; Macrì, F; Vianello, A. Activity of a K+ATP channel in Arum spadix mitochondria during thermogenesis. *Journal of Plant Physiology*, 2008b, 165, 1360-1369.

Petrussa, E; Casolo,V; Braidot, E; Chiandussi, E; Macrì, F; Vianello, A. Cyclosporin A induces the opening of a potassium-selective channel in higher plant mitochondria. *Journal of Bioenergetics and Biomembranes*, 2001, 33, 107-117.

Reintanz, B; Szyroki, A; Ivashikina, N; Ache, P; Godde, M; Becker, D; Palme, K; Hedrich, R. AtKC1, a silent Arabidopsis potassium channel α-subunit modulates root hair K+ influx. *Proceedings of the National Academy of Sciences of USA,* 2002, 99, 4079-4084.

Ruy, F; Vercesi, AE; Andrade, PBM; Bianconi, ML; Chaimovich, E; Kowaltowski, AJ. A highly active ATP-insensitive K+ import pathway in plant mitochondria. *Journal of Bioenergetics and Biomembranes*, 2004, 36, 195-202.

Shieh, C-C; Coghlan, M; Sullivan, JP; Gopalakrishnan, M. (2000). Potassium channels: Molecular defects, diseases, and therapeutic opportunities. *Pharmacological Reviews*, 2000, 52, 557-593.

Siemen, D; Loupatatzis, C; Borecky, J; Gulbins, E; Lang, F. Ca2+-activated K channel of the BK-type in the inner mitochondrial membrane of a human glioma cell line. *Biochemical and Biophysical Research Communications,* 1999, 257, 549-554.

Szabò, I; Bock, J; Jekle, A; Soddemann, M; Adams, C; Lang, F; Zoratti, M; Gulbins, E. A novel potassium channel in lymphocyte mitochondria. *The Journal of Biological Chemistry*, 2005, 280, 12790-12798.

Tanaka, N; Fujita, M; Handa, H; Murayama, S; Uemura, M; Kawamura, Y; Mitsui, T; Mikami, S; Tozawa, Y; Yoshinaga, T; Komatsu, S. Proteomics of the rice cell: systematic identification of the protein populations in subcellular compartments. *Molecular Genetics and Genomics*, 2004, 271, 566-576.

Trono, D; Flagella, Z; Laus, MN; Di Fonzo, N; Pastore, D. The uncoupling protein and the potassium channel are activated by hyperosmotic stress in mitochondria from durum wheat seedlings. *Plant, Cell and Environment*, 2004, 27, 437-448.

Yarov-Yarovoy, V; Paucek, P; Jaburek, M; Garlid, KD. The nucleotide regulatory sites on the mitochondrial K(ATP) channel face the cytosol. *Biochimica and Biophysica Acta,* 1997 , 1321, 128-136.

Zingman, LV; Alekseev, AE; Hodgson-Zingman, DM; Terzic, A. ATP-sensitive potassium channels: metabolic sensing and cardioprotection. *Journal of Applied Physiology*, 2007, 103, 1888-1893.

Zottini, M; Formentin, E; Scattolin, M; Carimi, F; Lo Schiavo, F; Terzi, M. Nitric oxide affects plant mitochondrial functionality in vivo. *FEBS Letters,* 2002 , 515, 75-78.

In: Potassium Channels
Editor: Danielle S. Fonseca

ISBN: 978-1-61324-880-5

Chapter VI

Involvement of the Mitochondrial ATP-Sensitive Potassium Channel in the Beneficial Effects of Fasting on the Ischemic-Reperfused Rat Heart*

M. G. Marina Prendes, M. S. González, R. Hermann, N. G. Pascale, M. E. Torresín, M. M. Jaitovich, E. A. Savino and A. Varela

Department of Biological Sciences, Faculty of Pharmacy and Biochemistry, University of Buenos Aires and IQUIMEFA-CONICET, Buenos Aires, Argentina

---

* A version of this chapter was also published in Handbook of Nutritional Biochemistry: Genomics, Metabolomics and Food Supply edited by Sondre Haugen and Simen Meijer, published by Nova Science Publishers, Inc. It was submitted for appropriate modifications in an effort to encourage wider dissemination of research.

Abstract

Fasting improves contractile recovery and attenuates lactate production and mitochondrial permeability transition (MPT) without altering cell viability in ischemic-reperfused rat hearts. 5-hydroxydecanoate (5-HD), a mitochondrial ATP-sensitive K^+ channel (K-ATP_{mito}) blocker, abolishes the improvement of mechanical recovery elicited by fasting despite it decreases lactate production, suggesting that K-ATP_{mito} might be involved in the preservation of contractility. This chapter assessed the contribution of K-ATP_{mito} in the attenuation of MPT which in turn may improve mitochondrial energetics.

Langendorff-perfused hearts from fed and fasted rats were subjected to ischemia-reperfusion in the presence or absence of 5-HD. To assess whether 5-HD has any direct effect on glycolysis, a cell free heart extract containing all the glycolytic enzymes was used. MPT was quantitated measuring the mitochondrial 2-[^{3}H]-deoxyglucose (^{3}H-2-DG) entrapment. Total heart ^{3}H-2-DG content as an estimation of necrosis was measured. As an index of mitochondrial energetic function, the rate of ATP synthesis was examined in mitochondria isolated from the ischemic-reperfused hearts.

Fasting increased the rate of mitochondrial ATP synthesis and attenuated MPT. 5-HD abolished these protective effects. 5-HD did not change total heart ^{3}H-2-DG content indicating that it lacked effects on cell survival. Since 5-HD did not affect glucose consumption and lactate production in the cell-free heart-extract the inhibition of glycolysis could be due to its own oxidative metabolism.

It may be concluded that cardioprotection elicited by fasting in the ischemic-reperfused heart could be ascribed, at least in part, to the K-ATP_{mito} activation which in turn inhibits MPT and induces energetic preservation.

Introduction

Previous fasting, which enhances triglyceride stores and accelerates oxidation of fatty acids derived from endogenous lipolysis [1-2] has been shown to improve contractile recovery, decrease the ischemic lactate production [1,3-5], lessen oxidative damage and attenuate mitochondrial permeability transition (MPT) [6] in the Langendorff perfused rat hearts subjected to no flow global ischemia and reperfusion. This fasting-induced cardioprotection occurred in spite of the previously reported ischemic increase of long-chain fatty-acyl CoA tissue levels [5,7] which have been reported to induce MPT, cytochrome c release and inhibition of state 3 respiration of isolated rat liver mitochondria [8]. Furthermore, fatty-acyl CoA esters inhibit the mitochondrial ATP-sensitive K^+ channel (K-ATP_{mito}) opening [9] to which has been ascribed a central role as a protector against ischemia-reperfusion injury as well as in the ischemic and pharmacological preconditioning [10-12].

Regarding the aforementioned findings, it is interesting to note that it has been reported that 5-hydroxydecanoate (5-HD), a powerful inhibitor of the K-ATP_{mito}, abolishes the improvement of post-ischemic mechanical recovery occurring in the fasted rat heart despite it decreases the ischemic lactate production [13], suggesting that K-ATP_{mito} is involved in the preservation of the contractile function in this nutritional state.

On these bases, the aim of the present chapter was to explore the contribution of K-ATP_{mito} in attenuation of MPT elicited by fasting in the ischemic-reperfused hearts. To meet this goal it was investigated whether 5-HD abrogates the inhibitory effect of fasting on the MPT using the "hot dog" technique whereby ^{3}H-2-DG is entrapped in mitochondria that have undergone MPT [14]. Present chapter also aimed to ascertain the effects of fasting and the involvement of K-ATP_{mito} in the preservation of mitochondrial bioenergetics assessed by measuring the rate of ATP synthesis in the mitochondria isolated from ischemic-reperfused hearts.

Furthermore, as it was mentioned above, 5-HD inhibits glycolysis in the ischemic hearts, therefore it was considered interesting to assess whether it has any direct effect on this metabolic pathway.

The data reported here argue for a role of K-ATP_{mito} activation and its impact on mitochondrial permeability and energetics preservation as a relevant factor in the recovery of contractile function of viable myocardium of the ischemic-reperfused fasted rat hearts.

Materials and Methodology

Experimental Protocol

The investigation conforms with the Guide for the Care and Use of Laboratory Animals published by the US National Institutes of Health (NIH Publication N° 85-23, revised 1996) and the Argentine Republic Law N° 14346 concerning animal protection. 58 Wistar female rats weighing 250-350 g, maintained on a 12-h dark/light cycle, fed *ad libitum* or fasted-24 h were used. Rats were anesthetized with diethylether and heparin (250 IU) was injected into the jugular vein. The hearts were quickly isolated and cooled in ice-cold saline until contractions stopped. The hearts were subsequently mounted on a modified Langendorff apparatus (Hugo Sachs Elektronik) and isovolumically perfused at a constant pressure of 70 mmHg with a non recirculating Krebs-Ringer bicarbonate solution of the following composition (mM): NaCl 120, $NaHCO_3$ 25, KCl 4.8, $MgSO_4$ 1.33, KPO_4H_2 1.2, $CaCl_2$ 1.6, Na_2EDTA 0.02 and glucose 10. The perfusate was gassed with 95% O_2-5% CO_2 (pH 7.4) and kept at a constant temperature of 37° C. After a 30 min equilibration period, the hearts were subjected to 25 min of no flow global ischemia and 30 min RP. Ischemia was begun by shutting off perfusate flow. 100 μM 5-HD was added into the perfusion medium 10 min before the onset of ischemia and remained all over the experiment. Only hearts with left ventricular developed pressure > 60 mm Hg and heart rate >200 beats/min at the end of the equilibration period were included.

Mitochondrial ATP Synthesis

Mitochondria were isolated from ischemic-reperfused hearts by differential centrifugation after tissue homogenization [15] in ice-cold sucrose buffer solution (300 mM sucrose, 10 mM Tris-Cl, 2 mM EGTA, 5 mg/ml BSA, pH 7.4). The mitochondrial pellet was then washed three times in sucrose isolation buffer solution lacking BSA. Cardiac mitochondria prepared with this procedure have been shown to be metabolically active with respiratory control ratios of 3.5 to 5.0 with succinate and 8.0 to 10.0 with glutamate/malate and corresponding ADP/O ratios of 1.5 to 1.7 and 2.5 to 2.7 [15]. Since it is well documented that complex I of the respiratory chain is most sensitive to reperfusion injury [16], mitochondrial ATP synthesis was measured in the presence of the complex I substrates pyruvate and malate. Mitochondria (1 mg protein/ml) were incubated for 10 min in medium containing (mM): KCl 125, Mops 20, Tris 10, EGTA 0.5, KH_2PO_4 2.4, Cl_2 Mg 2.5, malate 2.5, pyruvate 2.5, pH 7.4 in a 25° C metabolic shaker. After 2 minutes ATP synthesis was initiated by the addition of 2.5 mM ADP. ADP 2.5 mM was used because it corresponds to a physiological concentration found in myocytes.

Aliquots were taken from incubation at 3-min intervals for 10 min, and reactions were stopped by adding the samples to perchloric acid. The neutralized supernatant was assayed for ATP by luciferin-luciferase luminometry (Sigma bioluminescent assay kit). Mitochondrial protein concentration was determined by the method of Lowry using bovine serum albumin as a standard, and the rate of mitochondrial ATP synthesis was calculated and expressed as μmoles of ATP synthesis per minute per milligram of mitochondrial protein.

Assessment of MPT *In Situ* Using Mitochondrial ^{3}H-2-DG Entrapment

This was performed as described previously [14,17]. After a 15 min stabilization, hearts were perfused for 30 min in recirculating mode with 40 ml of Krebs-Ringer bicarbonate solution containing 0.5 mM ^{3}H-2-DG (0.1 μCi / ml). Perfusion was then returned to non recirculating mode with normal medium in presence or absence of 100 μM 5-HD and continued for 10 min before induction of ischemia as outlined above. This, washes out extracellular ^{3}H-2-DG from the heart while cytosolic ^{3}H-2-DG remains entrapped. At the end of reperfusion, ventricles were rapidly removed from the heart, weighed and homogenized in ice-cold sucrose buffer solution (300 mM sucrose, 10 mM Tris-Cl, 2 mM EGTA, 5 mg/ml BSA, pH 7.4). A sample of the total homogenate was retained for the measurement of total ^{3}H d.p.m. after protein precipitation by addition of an equal volume of 5% (w/v) $HClO_4$. The remainder of the homogenate was used to isolate mitochondria by using a 2 min centrifugation at 2000 x g to remove cell debris, followed by centrifugation of the supernatant at 10000 x g for 5 min to sediment the mitochondria. The mitochondria pellet was then washed three times in sucrose isolation buffer solution lacking BSA and resuspended in a final volume of 2.5 ml of isolation buffer solution. A 100 μl portion of mitochondrial suspension was retained for assay of citrate synthase, and 0.4 ml of 0.5 % $HClO_4$ was added to the remainder to release entrapped ^{3}H d.p.m. Protein was precipitated by centrifugation at 8000 x g for 2 min, and the resulting supernatant was counted for radioactivity in 2.5 ml of scintillant liquid (Ecolite, ICN). The mitochondrial uptake of ^{3}H-2-DG was calculated on the basis of ^{3}H d.p.m. measured in the mitochondrial extract. In order to correct for variation between experiments in loading of the hearts with ^{3}H-2-DG, total homogenate ^{3}H d.p.m. per g wet weight of tissue was determined. In order to correct for variations in the recovery of mitochondria between experiments, citrate synthase activity was measured spectrophotometrically [18] in the mitochondrial fraction, and the mitochondrial ^{3}H d.p.m. was expressed per unit of this enzyme. ^{3}H-2-DG uptake in ^{3}H d.p.m./unit of citrate synthase was expressed as a ratio with respect to the total tissue ^{3}H d.p.m. This ratio should be independent of both ^{3}H-2-DG cell loading and mitochondrial recovery. Mitochondria were prepared in the presence of EGTA to chelate calcium in order to attain a rapid closure of the pore and entrapment of the ^{3}H-2-DG within the matrix.

Measurement of the Effects of 5-HD on a Cell-Free Soluble Heart Extract Containing the Glycolytic Enzymes Present in the Heart Fibers

Direct effects of 5-HD, if any, on glucose catabolism and lactate accumulation were assessed using a soluble heart extract considered to contain all the glycolytic enzymes present in heart cells in order to avoid interference from other metabolic pathways on glycolytic flux.

The following procedure, derived from that of Wu and Davis [19] has been previously described by Varela and Savino [20]. Rats anesthetized with diethylether, heparin (250 IU) were injected into the jugular vein, and the hearts were rapidly removed. The heart ventricle was homogenized in 50 mM NaH_2PO_4, 100mM KCL, pH 7.2 and the final concentration of tissue in the homogenate was 200 mg/ml. The homogenate was centrifuged at 12000 x g during 10 min and then at 50000 x g for 60 min. The supernatant was placed in a dialyzing tube and concentrated with polyethylene glycol 6000, to approximately 19 mg protein per ml. The concentrated extract was placed on a Sephadex G-25 coarse column and eluted with the homogenization buffer to remove metabolites and cofactors. Then, 0.5 ml aliquots of this heart extract were incubated for 30 min with 0.5 ml 0.15 M KCl and 0.2 ml of the reaction mixture.

The final composition of the reaction medium was as follows: 7.8 mg protein/ml, 6 mM $MgSO_4$, 17 mM NaH_2PO_4, 96 mM KCl, 2.4 mM AMP, 3 mM ATP, 1 mM NAD^+, 10 mM nicotinamide, 3 mM glucose, pH 7.4. All handling was conducted at 4° C except for the incubations which were performed at 37° C for 30 min in a metabolic shaker. Incubations were performed both in absence and in presence of 100 μM 5-HD. Reactions were stopped by transferring 1.2 ml of the reaction media to a flask containing 1.8 ml of 0.66 mM $HClO_4$. Glucose and lactate were measured enzymatically. All enzymes, cofactors and metabolites were purchased from *Sigma*.

Statistical Analysis of Data

All values are expressed as means ± SE. ANOVA and the Tukey's post hoc test was used to determine whether any significant differences existed among groups. Significance was set at $P < 0.05$ level.

Results

As can be seen in Figure 1, the rate of ATP synthesis in mitochondria isolated from hearts exposed ischemia-reperfusion was about 51 % higher in the fasted with respect to the fed group. Addition of 5-HD into the perfusate did not affect the rate of ATP synthesis in the fed rat heart, but reduced it about 46% in the fasted rat hence abolishing the differences between both nutritional states.

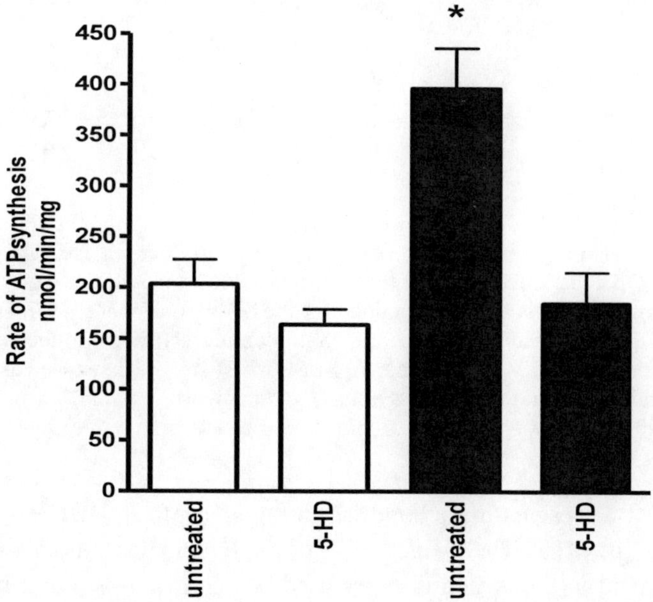

Figure 1. ADP-stimulated ATP production in isolated mitochondria from hearts subjected to ischemia-reperfusion in the presence (5-HD) or in the absence (untreated) of 5-hydroxydecanoate 100 mM. Open bars: fed hearts; filled bars: fasted hearts. Values are the mean±SEM (n=10) and are expressed as nmol / min / mg mitochondrial protein. *: p<0.01 vs untreated fed, 5-HD fed, 5-HD fasted.

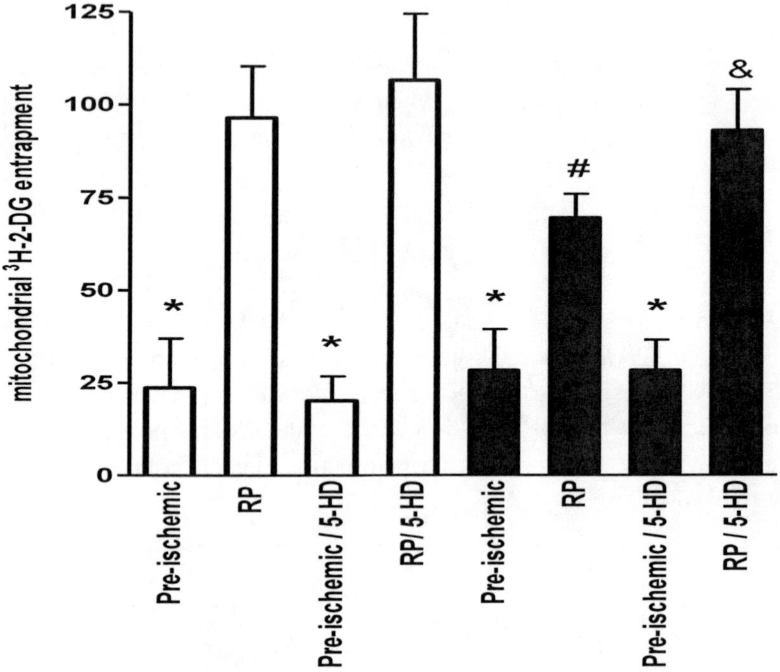

Figure 2. Effects of 5-hydroxydecanoate (5-HD; 100 mM) on mitochondrial permeability transition, measured *in situ* by 2-deoxy-^{3}H-glucose (^{3}H-2-DG) entrapment. Open bars: fed hearts; filled bars: fasted hearts. 5-HD was given 10 min before the onset of ischemia and remained for the duration of the experiment. Pre-ischemic refers to the end of preischemic 25 min period. RP refers to the end of the 30 min reperfusion period. Values are mean±SEM (n= 8) and are expressed as x 10^5 ratio mitochondrial ^{3}H d.p.m./unit citrate synthase to total heart ^{3}H d.p.m./g wet weight. *: $p< 0.01$ vs 30 min RP, #: $p<0.05$ vs 30min RP fed hearts, and: $p< 0.05$ vs 30 min RP fasted hearts.

Figure 2 represents mitochondrial entrapment of ^{3}H-2-DG as an estimate of the extent of MPT. Then, since 24-h fasting attenuated mitochondrial ^{3}H-2-DG entrapment in the ischemic-reperfused hearts, it is likely that it decreased MPT. Data here reported also indicate that mitochondrial entrapment of ^{3}H-2-DG was not affected by 5-HD in the fed hearts whereas it was significantly increased in hearts from fasted hearts, abolishing differences between both nutritional states. On the other hand, cell retention of ^{3}H-2-DG was similar in both nutritional groups either in the absence or the presence of 5-HD (Figure 3).

Furthermore, data shown in Figure 4 indicate that there were no significant differences between groups in the mitochondrial citrate synthase activity that was used as a quantitative enzyme-marker for the content of intact mitochondria.

Since previous data [13] showed that 5-HD lowered the lactate accumulation during the ischemic period and in order to avoid interference from other metabolic pathways on glycolytic flux, the effects of 5-HD on glucose disappearance and lactate accumulation were assessed using a soluble cell extract considered to contain all the glycolytic enzymes present in heart cells. Addition of 100 μM 5-HD into the heart extract did neither alter glucose utilization (1.70 ± 0.30 *vs* 2.07 ± 0.35 mmol / L per 30 min in the control and 5-HD groups, respectively; n=5) nor lactate accumulation (1.2 ± 0.4 *vs* 1.3 ± 0.4 mmol / L per 30 min in the control and 5-HD groups, respectively).

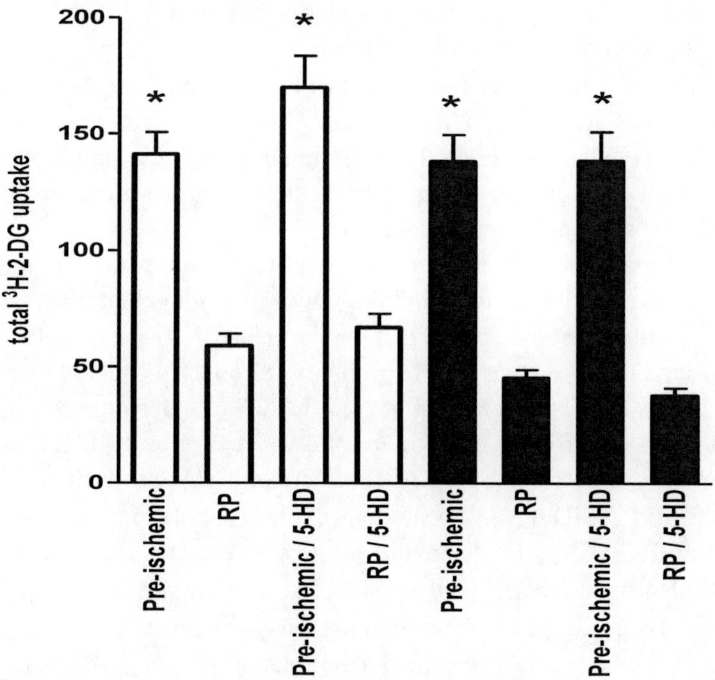

Figure 3. Effects of 5-hydroxydecanoate (5-HD; 100 mM) on total 2-deoxy-^{3}H-glucose (^{3}H-2-DG) uptake. Open bars: fed hearts; filled bars: fasted hearts. 5-HD was given 10 min before the onset of ischemia and remained for the duration of the experiment. Pre-ischemic refers to the end of preischemic 25 min period. RP refers to the end of the 30 min reperfusion period. Values are mean±SEM (n= 8) and are expressed as total ^{3}H-2-DG uptake x 103 d.p.m./g wet weight. *: p< 0.01 vs 30 min RP.

Conclusion

Marina Prendes et al. [6] have demonstrated that functional protection elicited by previous fasting to the ischemic-reperfused heart is accompanied by attenuation of MPT. Present chapter broadened the understanding of the mechanisms involved in mitochondrial protection by the assessment of the role of mitochondrial oxidative phosphorilation. They also show that the $K\text{-}ATP_{mito}$ antagonist 5-HD which has been shown to reverse the improvement of functional recovery elicited by fasting [13] also abolished the protection afforded by fasting on MPT and mitochondrial ATP synthesis. Furthermore, the exposure of the ischemic-reperfused hearts of fed rats to 5-HD did not affect MPT and agreeing with previous data [21] neither affected the rate of ATP synthesis. These findings suggest the existence of a link between fasting and $K\text{-}ATP_{mito}$ activation which could be implicated in some protective mechanisms occurring at mitochondrial level. Supporting this proposal Korge et al. [22] and Costa et al. [12] have shown that activation of $K\text{-}ATP_{mito}$ results in inhibition of MPT and Ozcan et al. [23] demonstrated that $K\text{-}ATP_{mito}$ openers preserved mitochondrial respiration and ATP synthesis and limited mitochondrial damage, preventing cytochrome c release upon reoxygenation, effects that were antagonized by 5-HD.

Induction of the inner membrane depolarization which in turn would reduce the driving force for calcium uptake by mitochondria and thereby preventing mitochondrial matrix calcium overload [22] may explain how $K\text{-}ATP_{mito}$ activation reduces MPT probably through its electrogenic effect. Although the link between fasting and $K\text{-}ATP_{mito}$ opening remains to be established, it is worthy to mention that it was demonstrated that inhibition of nitric oxide synthase with L-NAME decreased the protection elicited by fasting as did 5-HD [24]. Furthermore, nitric oxide donors facilitate the activation of $K\text{-}ATP_{mito}$ by diazoxide and in part directly activate this channel [25] and also have been shown to modulate the sensitivity of $K\text{-}ATP_{mito}$ to ATP [26]. Taken together, these findings suggest that nitric oxide might be involved in the beneficial effects of fasting acting on the $K\text{-}ATP_{mito}$, although further studies are necessary to resolve this hypothesis.

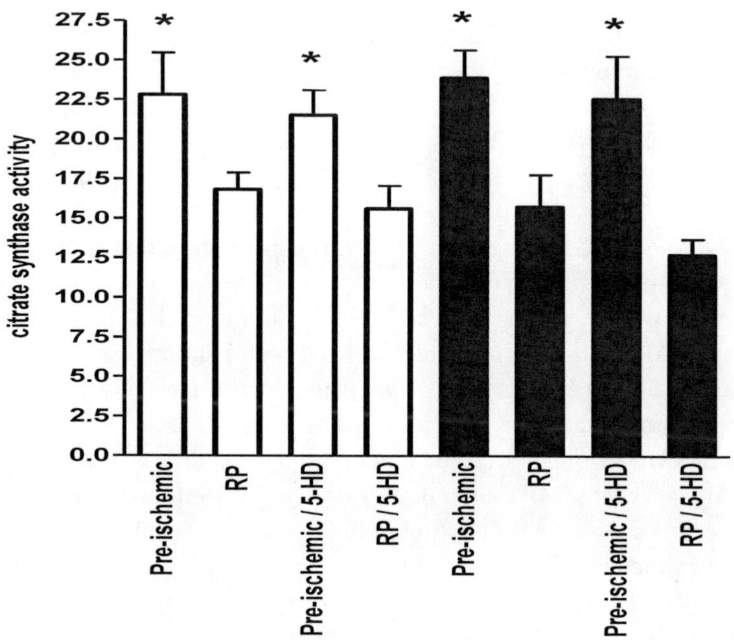

Figure 4. Effects of 5-hydroxydecanoate (5-HD; 100 mM) on mitochondrial citrate synthase activity. Open bars: fed hearts; filled bars: fasted hearts. 5-HD was given 10 min before the onset of ischemia and remained for the duration of the experiment. Pre-ischemic refers to the end of preischemic 25 min period. RP refers to the end of the 30 min reperfusion period. Values are mean±SEM (n= 8) and are expressed as citrate synthase units / g wet weight.*: $p< 0.05$ vs 30 min RP.

Since other effects of 5-HD unrelated to K-ATP_{mito} have been described [27-30], other potentially detrimental mechanisms might contribute to the inhibitory effects of this K-ATP_{mito} blocker on the protection elicited by fasting. It has been shown that 5-HD serves as substrate for acyl-CoA synthetase [27,31] and 5-hydroxydecanoyl-CoA is transported into mitochondria and metabolized via the ß-oxidation pathway, albeit with rate limiting kinetics at the penultimate step, creating a bottleneck for ß-oxidation of fatty acids [29]. Interestingly, state-3 respiration of heart mitochondria induced by the physiological substrates decanoyl-CoA or lauryl-CoA was reduced by 40 % after addition of 100 μM 5-hydroxydecanoyl-CoA [29], suggesting that 5-HD acts as inhibitor of fatty acid oxidation in heart mitochondria. Moreover, mitochondria isolated from 5-HD treated ischemic-reperfused hearts show increases in matrix volume and inhibition of respiration [31]. It should also be pointed out that ischemic-reperfused rat hearts exhibited lower lactate ischemic production in spite of higher preischemic glycogen content in the presence of 5-HD in both nutritional states [13] indicating that it limited glycolytic flux. Since present data showed that 5-HD lacks direct inhibitory effects on the glycolytic rate of rat hearts it cannot be ruled out that inhibition of glycolysis was due to its own metabolism.

Regarding these findings it is worthwhile to mention that exogenous short- and medium-chain fatty acids have been previously reported to inhibit the heart glycolytic flux during oxygen limited conditions [1,32]. Furthermore, 4-pentenoic acid which, reminiscently to 5-HD inhibits fatty acid oxidation but can also be ß-oxidized through a pathway different to that involved in its inhibitory properties [33,34], reduces hypoxic-lactate production [32]. However, 5-HD abolished the beneficial effects of fasting but did not affect contractile recovery, MPT or mitochondrial ATP synthesis of the fed group despite it limited lactate production in both nutritional states. Therefore, these cluster of evidence indicates that the inhibition of glycolysis is unrelated to the blockade of K-ATP_{mito} whereas the noxious effects of 5-HD on contractile function in the fasted hearts indeed appear to be linked to the K-ATP_{mito}-blockade.

Finally, in accordance with previous data showing that fasting had no effects in cell viability preservation [13] and CK leakage [5-6] in the hearts exposed to ischemia-reperfusion, present data show an almost equal cellular retention of ^{3}H-2-DG together with similar activity of CS, the marker enzyme for the inner mitochondrial membrane, in both nutritional groups. Furthermore, coinciding with the previously reported lack of effects of 5-HD on cell viability [13]in the present chapter it was also found that there were no differences in the cellular retention of ^{3}H-2-DG together with a similar activity of citrate synthase either in the absence or the presence of this inhibitor. These data indicate that mitochondria from necrotic cells would be totally disrupted so they could not be collected in the mitochondrial pellet. Thus, findings reported here, suggest that most likely there is a similar content of intact mitochondria in a similar number of intact surviving cells, although they exhibit less MPT and improved ATP synthesis in the fasted group not exposed to 5-HD. Furthermore, since all ^{3}H-2-DG that might have been entrapped in the cytosol and mitochondria will be lost and not detected in necrotic cells, ^{3}H-2-DG technique does detect the cells that have not yet progressed to necrosis but are perhaps on their way and exhibit compromised mitochondrial function associated with MPT [35]. Since MPT results in the collapse of the mitochondrial membrane potential and the impairment of ATP synthesis, a smaller MPT and consequently a greater ATP production in fasted rat mitochondria might effect an improvement of contractile function in still viable myocardium. However, the increase in ATP synthesis might not be enough to protect the cardiomyocytes of the fasted rat against necrosis.

In summary, it may be concluded that most likely the noxious effects of 5-HD on the ischemic-reperfused fasted-rat heart are the consequence of K-ATP_{mito} blockade aside from its involvement in the heart fatty acid and glucose metabolism. Therefore, the cardioprotection elicited by fasting seems to be caused, at least in part, through the K-ATP_{mito} activation and its consequences on mitochondrial permeability and energetic preservation.

Acknowledgments

The authors thank Norma Gladys Infante for excellent technical assistance.

Research supported by grants from Universidad de Buenos Aires, IQUIMEFA-CONICET and ANPCyT BID contract 1728 / OC-AR project Nº 26026.

References

[1] Montessuit, C., Papageorgiou, I., Tardy, I. and Lerch, R. (1996). Effect of nutritional stage on substrate metabolism and contractile function in postischemic rat myocardium. *Am. J. Physiol.*, *271*, H2060-70.

[2] Varela, A. and Savino, E. A. (1988). Endogenous triacylglycerol utilization by the isolated rat atria. *Rev. esp. Fisiol.*, *44*, 87-92.

[3] Schneider, C. A. and Taegymeyer, H. (1991). Fasting in vivo delays myocardial cell damage after brief periods of ischemia in the isolated working rat hearts. *Circ. Res.*, *68*, 1045-50.

[4] Montessuit, C., Papageorgiou, I., Tardy-Cantalupi, I., Rosenblatt-Velin, N. and Lerch, R. J. (2000). Postischemic recovery of heart metabolism and function: role of mitochondrial fatty acid transfer. *J.Appl. Physiol.*, *89*, 111-9.

[5] Varela, A., Marina Prendes, M.G., Testoni, G., Vázquez, N., Astudilla, C., Cerruti, S. and Savino, E. A. (2002). Influence of fasting on the

effects of ischemic preconditioning in the ischemic-reperfused rat heart. *Arch. Physiol. Biochem.*, *110*, 189-96.

[6] Marina Prendes, M. G., González, M. S., Hermann, R., Pascale, N. G., Torresín, M. E., Jaitovich, M. M., Savino, E. A. and Varela, A. (2009). Involvement of mitochondrial permeability transition, glutathione status, pentose phosphate pathway and oxidative damage in the protective effect of fasting on the ischemic-reperfused rat heart. *Clin. Exp. Pharmacl. Physiol.*, (in press).

[7] Idell-Wenger, J. A., Grotoyohann, L. W. and Neely, J. R. (1978) Coenzyme A and carnitine distribution in normal and ischemic hearts. *J. Biol. Chem.*, *253*, 4310-8.

[8] Furuno, T., Kanno, T., Arita, K., Asami, M., Utsumi, T., Doi, Y., Inoue, M. and Utsumi, K. (2001). Roles of long chain fatty acids and carnitine in mitochondrial membrane permeability transition. *Biochem. Pharmacol.*, *62*, 1037-46.

[9] Paucek, P., Yarov-Yarovoy, V., Sun, X. and Garlid, K. D. (1996). Inhibition of mitochondrial KATP channel by long-chain acyl-CoA esters and activation by guanine nucleotides. *J. Biol. Chem.*, *271*, 32084-8.

[10] Facundo, H. T. F., Fornazari, M. and Kowaltowski, A. J. (2006). Tissue protection mediated by mitochondrial K channels. *Biochim. Biophys. Acta*, *1762*, 202-12.

[11] Szewczyk, A., Skalska, J., Glab, M., Kulawiak, B., Malinska, D., Koszela-Piotrowska, I. and Kunz, W. S. (2006). Mitochondrial potassium channels: From pharmacology to function. *Biochim. Biophys. Acta.*, *1757*, 715-20.

[12] Costa, A. D. T., Jakob, R., Costa, C. L., Andrukhiv, K., West, I. C. and Garlid, K. D. (2006). The mechanism by which the mitochondrial ATP-sensitive K+ channel opening and H_2O_2 inhibit the mitochondrial permeability transition. *J. Biol. Chem.*, *281*, 20801-8.

[13] Marina Prendes, M. G., García, J. V., Fernández, A., Pérez, M. J., Perazzo, J. C., Savino, E. A. and Varela, A. (2005). Effects of 5-hydroxydecanoate and ischemic preconditioning on the ischemic-reperfused heart of fed and fasted rats. *J. Physiol. Biochem.*, *61*, 447-50.

[14] Griffiths, E. J. and Halestrap, A. P. (1995). Mitochondrial non-specific pores remain closed during cardiac ischemia, but open upon reperfusion. *Biochem. J.*, *307*, 93-8.

[15] Solem, L. and Wallace, K. (1993). Selective activation of the sodium-independent, cyclosporin A sensitive calcium pore of cardiac mitochondria by doxorubicin. *Toxicol. Appl. Pharmacol.*, *121*, 150-7.

[16] Solani, G. and Harris, D. A. (2005). Biochemical dysfunction in heart mitochondria exposed to ischaemia and reperfusion. *Biochem. J.*, *390*, 377-94.

[17] Marina Prendes, M. G., Torresín, E., González, M., Fernández, M. A., Perazzo, J. C., Savino, E. A. and Varela, A. (2008). Protection of ischemic-reperfused rat heart by dimethylamiloride is associated with inhibition of mitochondrial permeability transition. *Clin. Exp. Pharmacol. Physiol.*, *35*, 201-6.

[18] Shepherd, D. and Garland, P. B. (1969). The kinetic properties of citrate synthase from rat liver mitochondria. *Biochem. J.*, *114*, 597-610.

[19] Wu, T. F. L. and Davis, E. J. (1981). Regulation of glycolytic flux in an energetically controlled cell-free system: the effects of adenine nucleotide ratios, inorganic phosphate, pH and citrate. *Arch. Biochem. Biophys.*, *209*, 85-99.

[20] Varela, A. and Savino, E. A. (1987). Effects of long chain fatty-acyl esters of coenzyme A and carnitine on cell-free rat heart preparations. *Arch. Int. Physiol. Biochim.*, *95*, 91-95.

[21] Fryer, R. M., Eells, J. T., Hsu, A. K., Henry, M. M. and Gross, G. J. (2000). Ischemic preconditioning in rats: role of mitochondrial KATP channel in preservation of mitochondrial function. *Am. J. Physiol.*, *278*, H305-12

[22] Korge, P., Honda, H. M. and Weiss, J. N. (2002). Protection of cardiac mitochondria by diazoxide and protein kinase C: Implications for ischemic preconditioning. *Proc. Natl. Acad. Sci.*, *299*, 3312-7.

[23] Ozcan, C., Bienengraeber, M., Dzeja, P, P. and Terzic, A. (2002). Potassium channel openers protect cardiac mitochondria by attenuating oxidant stress at reoxygenation. *Am. J. Physiol.*, *545*, *82*, H531-9.

[24] Marina Prendes, M. G., Gonzalez, M., Perez, J., Fernandez, A., Savino, E. and Varela, A. (2005). El L-NAME revierte parcialmente los efectos beneficiosos del diazoxido y del ayuno en corazones de rata sometidos a isquemia global-reperfusion. *Medicina*, *65(II)*, 95.

[25] Sasaki, N., Sato, T., Ohler, A., O'Rourke, B. and Marban, E. (2000). Activation of mitochondrial ATP-dependent potassium channels by nitric oxide. *Circulation*, *101*, 439-45.

[26] Shinbo, A. and Iijima, T. (1997). Potentiation by nitric oxide of the ATP-sensitive K+ current induced by K+ channel openers in guinea-pig ventricular cells. *Br. J. Pharmacol.*, *120*, 1568-74.

[27] Hanley, P. J., Mickel, M., Loffler, M., Brandt, U. and Daut, J. (2002). KATP channel-independent targets of diazoxide and 5-hydroxydecanoate in the heart. *J. Physiol.*, *542*, 735-45.

[28] Hanley, P. J., Gopalan, K. V., Lareau, R. A., Srivastava, D. K., von Meltzer, M. and Daut, J. (2003). β-oxidation of 5.hydroxydecanoate, a putative blocker of mitochondrial ATP-sensitive potassium channels. *J. Physiol.*, *547*, 387-93.

[29] Hanley, P. J., Drose, S., Brandt, U., Lareau, R. A., Banerjee, A. L., Srivastava, D. K., Banaszak, L. J., Barycki, J. J., Van Veldhoven, P. P. and Daut, J. (2005). 5-Hydroxydecanoate is metabolized in mitochondria and creates a rate limiting bottleneck for s-oxidation of fatty acids *J. Physiol.*, *562*, 307-18.

[30] Dzeja, P. P., Bast, P., Ozcan, C., Valverde, A., Holmuhamedov, E. L., Van Wylen, D. G. L. and Terzic, A. (2003). Targeting nucleotide-requiring enzymes: implications for diazoxide-induced cardioprotection. *Am. J. Physiol.*, *284*, H1048-56.

[31] Lim, K. H. H., Javadov, S. A., Das, M., Clarke, S. J. and Saadeh Suleiman, M. (2003). The effects of ischaemic preconditioning, diazoxide and 5-hydroxydecanoate on rat heart mitochondrial volume and respiration. *J. Physiol.*, *45*, 961-74.

[32] Varela, A., Lanzetta, D. and Savino, E. A. (1989). The effects of 4-pentenoic and pentanoic acids on the hypoxic rat atria. *Arch. Int. Physiol. Biochim.*, *97*, 375-80.

[33] Bremer, J. (1983). Carnitine-Metabolism and functions. *Physiol. Rev.*, *63*, 1420-80.

[34] Varela, A., Felip, M. N., Montesi, A. and Savino, E. A. (1988). The effects of 4-pentenoic and pentanoic acids on the isolated rat atria. *Arch. Int. Physiol. Biochim.*, *96*, 165-70.

[35] Halestrap, A. P., Clarke, S. J. and Javadov, S. A. (2004). Mitochondrial permeability transition pore opening during myocardial reperfusion. *Cardiovasc. Res.*, *61*, 372-85.

Index

A

D

E

F

G

H

I

K

L

M

N

O

P

Q

R

S

T

U

V

W

X

Z